W9-CQP-772

AN INTRODUCTION TO

CALIFORNIA PLANT LIFE

by

ROBERT ORNDUFF

UNIVERSITY OF CALIFORNIA PRESS
BERKELEY · LOS ANGELES · LONDON

CALIFORNIA NATURAL HISTORY GUIDES
Arthur C. Smith, General Editor

ACKNOWLEDGMENTS

Photographic credits: Spencer C. H. Barrett (Pl. 4A; 5A, C; 6A; 7A; 8D; 9A, B; 11D; 15D), Rolf Benseler (Pl. 2D), Sherwin Carlquist (Pl. 1B; 2B), David Hafner (Pl. 9D), L. Maynard Moe (Pl. 5B; 7B; 12C, D; 13C; 15B; 16B, D) Arthur C. Smith (Pl. 10C, D; 11A; 13A, B; 14D; 15A), Jepson Herbarium (Pl. 1C; 4D; 6C; 7C, D; 8A; 14B, C; 15C). Other photographs are by the author.

University of California Press
Berkeley and Los Angeles, California
University of California Press, Ltd.
London, England

CONTENTS

INTRODUCTION

To many Californians, the wealth of the state lies in its gold, its petroleum, its timber, or its fertile valleys. To those of us who are amateur or professional botanists, or who simply enjoy "plant watching," the riches of California are also reflected in the diversity of wildflowers, shrubs, and trees that occur throughout the state. California is isolated from the rest of North America by deserts or mountains that have allowed the development within its boundaries of one of the most varied floras that occurs anywhere on earth. The plant cover ranges from the forests of the northern coast and the mountain slopes, to the woodlands and scrublands of the foothills and deserts, to the grasslands of the valleys. Our plants range in size from the stately Coast Redwoods of the fog-shrouded coast to the minute belly plants of the southern deserts, and in age from the venerable four-thousand-year-old Bristlecone Pines to the diminutive ephemeral annuals whose life span can be counted in weeks.

This small book introduces you to the plant life of California and tells you something of how plants are grouped into communities and what environmental influences determine the pattern of distribution of these communities in the state. It also discusses the origin of our flora, how plants are adapted to the diverse climates of the state, and how they respond to forest and chaparral fires, to unusual soils, to man, and to each other.

The contents of this book are adapted from a syllabus that I wrote for an Independent Study course developed for the University of California, Berkeley. Because most of my botanizing has been in northern California, the contents of the book are perhaps unevenly weighted toward this portion of the state. Drafts of portions of the manuscript were read by L. R. Heckard, J. R. McBride, D. R. Parnell, and a few other friends and colleagues to whom I am indebted for helpful suggestions. I am also grateful to S. C. H. Barrett, R. Benseler, S. Carlquist, D. Hafner, L. M. Moe, and A. C. Smith for allowing me to reproduce their slides, to C. Mentges for executing the line drawings, and to P. Watters for her assistance in preparing the manuscript. Several of the color illustrations have been taken from the slide collection of the Jepson Herbarium, University of California, Berkeley. I am also indebted to my students, who have proven time and again that there is still a great deal that I have to learn about California plants, and that most of this will be gained by studying the plants themselves and not what is written about them in books.

Map 1. Major topographical features of California. The portion of the state in the California Floristic Province is to the coastward side of the hatched line.

1. PLANT NAMES

Binomials and Common Names

By a long-standing general agreement among botanists, every plant has two names. One of these is the *genus* name (or, as an adjective, the generic name) and the other is the *species,* or specific, name. It is easy to remember which is which because of the similarity of the word generic to "general" and because the word specific means just what it says. A large number of plants also have *common* or vernacular names. Such names are familiar to a larger number of people than are generic and specific names. However, there are some serious defects in calling plants only by their common names. One difficulty is that a common name applied to a plant in one area may apply to another plant in another area. For example, White Pine in the western United States refers to *Pinus monticola;* in the eastern United States it refers to *Pinus strobus.* In the Pacific Northwest, the name Skunk Cabbage refers to *Lysichiton americanum,* but in California this name is often applied to species of the unrelated genus *Veratrum.* In short, one reason to be wary of common names is because one common name may apply to two or more very different plants and obviously there can be some serious problems in communication when only common names are used.

A second objection to the use of common names for plants is that one plant may have many common names (although it has only one generic and specific name). In California, the names California Bay, California Laurel, or even Pepperwood are unambiguous and there is little difficulty in using any one of them for *Umbellularia californica.* Once you cross over the border into southwestern Oregon, however, this evergreen tree changes its name to the Oregon Myrtle. I doubt whether most southern Oregonians would have the foggiest notion of what tree you meant by bay or laurel, but if you mentioned myrtle they would immediately know. I might also mention in passing

3

that the oft-repeated story that the "Oregon Myrtle" grows only in Oregon and the Holy Land is untrue unless you assume that the latter designation refers to California. Perhaps the extreme example of multiple common names is the number of common names applied to Douglas Fir (*Pseudotsuga menziesii*), which is an exceptionally important timber tree in western North America. At least 26 different common names are in use for this coniferous tree. One might expect such situations to exist for economically useful, widespread, and conspicuous species, but communication is not enhanced by having to remember 26 different names for one tree when it has a pair of names that are unambiguous.

Therefore, every plant has two names in the system that is called *binomial* (two-name) *nomenclature.* A genus may include as few as one species (as is the case of *Umbellularia,* which consists of *U. californica* only) or it may include several species (as does *Pinus* – *P. ponderosa, P. jeffreyi, P. contorta, P. radiata, P. coulteri,* etc.). The generic name is always capitalized and is generally treated as a Greek or Latin noun, even though the word may have originated from Japanese (*Tsuga*), Cherokee (*Sequoia*), or other languages. Such a name has gender, that is, it is masculine, feminine, or neuter. This is not particularly important to remember for present purposes, except to say that the species name of a plant must agree in gender with the genus. Species names of plants generally are not capitalized. Among plants, some species names, such as those derived from a person's name (like *douglasii, fremontii*) may be capitalized, but this is not necessary and perhaps for the sake of consistency, the best thing is not to capitalize any specific name.

Specific names are generally treated as Latin or Latinized names; sometimes this leads to curious consequences as will be seen below. Some examples of generic and specific names:

COMMON NAME	GENUS	SPECIES
Coast Redwood	*Sequoia*	*sempervirens*
Ponderosa Pine	*Pinus*	*ponderosa*
Ocotillo	*Fouquieria*	*splendens*
Joshua Tree	*Yucca*	*brevifolia*
Basin Sagebrush	*Artemisia*	*tridentata*
California Poppy	*Eschscholzia*	*californica*

4

You will probably find that pronouncing these binomials gives you some difficulty. The best advice that I can give on pronouncing names is to listen to an "expert" and use his pronunciation; your "expert" may mispronounce the words, but at least he tries, and this is the first thing to do. That is, pronounce the names out loud — to yourself if need be — and say them in a comfortable euphonious way. Chances are that you will therefore pronounce the words correctly; if not, at least you will be understood, and if you are off-base, then your listener may correct you (if he is more experienced than you). The pronunciations of many generic names of plants are given in Webster's Unabridged Dictionary, or other books such as E. C. Jaeger's *Source Book of Biological Names and Terms.* It is easy to be intimidated by generic and specific names, but remember that only a very small proportion of the 290,000 species of flowering plants on this earth have common names. Therefore, for communication purposes, or for identification purposes, the only name that exists for most plants is a binomial. Also, remember that most people can pronounce *Rhododendron, Eucalyptus,* and *Chrysanthemum* correctly, and these generic names are probably more complex than are the majority of the generic names in the California flora; at least they are no more difficult. The nomenclature used in this book mostly follows that of P. A. Munz in his *A California Flora* (1959) and *Supplement* (1968) which are now available combined in a single volume. Munz also provides accent marks to aid in the pronunciation of family, generic, and specific names.

The Meaning of Plant Names

What do the binomials mean? How are they derived? There are several answers to these questions. Specific names of plants may be taken from several sources. Often, but by no means always, they tell you something about the plant. Following are some examples of specific names for some native plants of California:

Aesculus californica (California Buckeye). "Californica" means "Californian." One might suspect that since other buckeye species are found in the Old World and in North America, whoever named this species was giving it a geographical designation to distinguish it from its relatives elsewhere.

Fragaria chiloensis (Coast Strawberry). "Chiloensis" means "from Chiloe," an island off the coast of Chile. The "-ensis" suffix means that plants are from whatever place is designated by the prefix. We have species in other genera named *idahoensis, utahensis, canadensis,* etc. *Fragaria chiloensis* is a native of both California and Chile; there are several other plant species with this particular distributional pattern.

Pinus monticola (Western White Pine). "Monticola" means "living in the mountains." There are several other pines in the California mountains, so this specific name is not as appropriate as it might be. Nevertheless, there is only one pine named *Pinus monticola.*

Sequoia sempervirens (Coast Redwood). "Sempervirens" means "evergreen." Since most conifers are evergreen, you might ask how this came to be attached to our redwood. When it was first named, the Coast Redwood was assigned to *Taxodium,* the Bald Cypress genus. *Taxodium* species are all deciduous, that is, they lose their leaves seasonally, usually in the autumn. Had the Coast Redwood really been a *Taxodium,* it would have been an unusual one in its evergreen characteristics. But it isn't a "bald" cypress. Later, the plant was transferred to a new genus, *Sequoia,* but because of the international code governing nomenclature of plants, the specific name carried across from one genus to the other. Thus, *Taxodium sempervirens* became *Sequoia sempervirens.*

Pinus albicaulis (Whitebark Pine). Here is a case of a common name paralleling the binomial. "Albicaulis" means "white stem," and refers to the whitish bark of this montane pine.

Pinus edulis (Pinyon Pine). "Edulis" means "edible," in reference to the edible "pine nuts" gathered from this tree.

Pinus torreyana (Torrey Pine). Named after John Torrey, an important 19th century American botanist who resided in New

York. Torrey named a number of California plant species (although the Torrey Pine was named in his honor by someone else). There is also a genus *Torreya* (California Nutmeg) in California.

Generic names also follow the general descriptive pattern discussed above, with a few exceptions. Generic names such as *Osmorhiza* (Sweet Cicely) means "odorous root" in reference to the fragrance of the crushed root of this relative of the carrot; *Lithocarpus* (Tan Oak) means "stone fruit" in Greek, an allusion to the hard acorns which, however, are probably no harder than acorns of the true oaks (*Quercus* spp.). *Rhododendron* means "red tree" in Greek, and this generic name was probably chosen because the first species described in this large genus has red flowers. Many genera also bear commemorative names: *Jepsonia* is named after an early professor of botany at the University of California, Berkeley, and author of the *Manual of the Flowering Plants of California*. *Eschscholzia* is named after a 19th century Russian explorer; *Munzothamnus* is named after P. A. Munz, author of *A California Flora*. *Lewisia* and *Clarkia* are named after the pair of early 19th century explorers of the American northwest; *Rafinesquea* and *Schmaltzia* are both named after the eccentric botanist of the 19th century, Constantine Rafinesque-Schmaltz. Some generic names have geographical connotations, e.g., *Hesperolinon* means "western flax" and refers to a group of species in a western genus closely related to the widespread flax genus *Linum*. Many generic names, particularly those of Old World plants, are taken from mythology, such as *Cassiope, Adonis,* and *Phoenix.*

Still other generic names refer to supposed medicinal properties of the plants: *Scrophularia* was named because of its use in treating scrofula; *Salvia* comes from the Latin *salveo,* "I save", in reference to the purported lifesaving abilities of this plant. Other names are from traditional usages that pre-date scientific botany; in essence, these are common names which have come into scientific usage. Among these are *Acer* (maples), *Quercus* (oaks), and *Pinus* (pines). A final category could perhaps be called whimsical names, which suggest that some taxonomists have a sense of humor, or at least a certain amount of ingenuity in devising taxonomic names. *Muilla* is a western plant that looks like an onion (*Allium*) although it differs from onions in that it is odorless, among other characters. *Muilla* is *Allium* spelled backwards. *Tellima* (fringe cups) is a western saxifrage whose name is an anagram of that of another genus in the same family: *Mitella* (mitrewort). Perhaps the ultimate is the curious generic name for a rare California aquatic plant named *Legenere limosa.* Its generic name is an anagram derived from the letters of the name E. L. Greene, another early professor of botany at the University of California, Berkeley, and a noted, indeed controversial, figure in the botanical history of the state.

There are also some generic and specific names which are truly misnomers, because of historical accident. Goatnut or Jojoba, a desert shrub of the American Southwest, otherwise known as *Simmondsia chinensis,* is not found in China. Apparently the specimen upon which the name is based was involved in a mixup of labels with a group of plants that had indeed been collected in China. One wonders how many Chinese plants are called "californica" as a consequence of this accident! Nevertheless, inappropriate as it is, the specific name of this shrub must remain. Thimble Berry, *Rubus parviflorus,* has a specific name which means "small flowered", yet it has one of the largest flowers of any member of the genus *Rubus* (which includes blackberries, blackcaps, etc.). How this misnomer came to be applied to Thimble Berry is uncertain, but it is possible that the specimen upon which the name was based was atypical in some respect. Lastly, those of you who have travelled up the northern California coast in Mendocino County may have visited the Mendocino White Plains, where there are several dwarfed conifers, including Pygmy Cypress, *Cupressus pygmaea.* Although this tree is truly a small one when it grows on the peculiar soils in these areas, when it occurs off these soils it grows to a good-sized tree and is hardly a pygmy.

The Taxonomic Hierarchy

Every species belongs to a genus. There is only one *Pinus lambertiana,* although there are many other species of pines, such as *Pinus torreyana, P. contorta, P. ponderosa, P. aristata,* and so on. (Note that once a generic name is used in a paragraph, it may be abbreviated by a single letter in subsequent usage. Specific names are not abbreviated, however.) The taxonomic hierarchy continues upward, since every genus belongs to a family. For example, the rose genus *Rosa,* the blackberry genus *Rubus,* the strawberry genus *Fragaria,* the bitterbrush genus *Purshia,* chokecherries (*Prunus*), cinquefoils (*Potentilla*), and their generic relatives all are members of the rose family (Rosaceae). Despite the differences in the general appearance of these plants, close examination of the flowers reveals that they have a number of basic similarities which indicates that they are related and should be placed together in a single family. Likewise, pines (*Pinus*), true firs (*Abies*), hemlocks (*Tsuga*), and a number of other coniferous tree genera are placed together in the Pinaceae.

The family name, like the generic name, is always capitalized. With a few exceptions family names terminate with the suffix "-aceae"; exceptions are the family names of some common families such as the grass family (Gramineae), sunflower family (Compositae), pea family (Leguminosae), and a few others. For the sake of consistency, those families that do not have the "-aceae" ending can be given other names with the "-aceae" ending:

Compositae	=	Asteraceae
Gramineae	=	Poaceae
Leguminosae	=	Fabaceae
Labiatae	=	Lamiaceae
Umbelliferae	=	Apiaceae

Either usage is acceptable.

In one respect, the taxonomic hierarchy is like an accordion in that various categories can be inserted at appropriate levels by using the prefix "sub." Subspecies are in frequent usage; subgenera and subfamilies are of less common occurrence in botanical nomenclature. For example, one of the perennial composites belonging to the goldfield genus *Lasthenia* has three subspecies. The most widespread of these is *L. macrantha* subsp. *macrantha,* which occurs in a couple of localities on the California

9

coast south of San Francisco Bay; it is common from Point Reyes northward into Mendocino County. A second subspecies is *L. macrantha* subsp. *bakeri* (named after Milo Baker, an important botanical explorer of the North Coast Ranges), which is restricted to shaded localities that occur slightly inland from those occupied by subsp. *macrantha*. The third subspecies is *L. macrantha* subsp. *prisca*, which is restricted to a few coastal headlands in southern Oregon.

The following chart will give you an idea of the hierarchical nature of the taxonomic system:

TAXONOMIC UNIT	CHARACTERS MEMBERS HAVE IN COMMON	TAXONOMIC CATEGORY
Pinus lambertiana (Sugar Pine)	1. Leaves with 1 vascular strand 2. Five leaves per cluster 3. Cones unarmed 4. Cones 10-16 in. (25.4-40.6 cm) long 5. Leaves sharp-pointed	Species
Pinus	1. Leaves of two kinds 2. Cones maturing after first year 3. Cone scales with minute bracts	Genus
Pinaceae	1. Cone scales overlapping 2. Cone scales with 2 seeds	Family
Coniferales	1. Cone dry, with several scales 2. 1 to several seeds per cone 3. Leaves needle-like (or scale-like)	Order

Using the characteristics in the preceding chart, one can guess that although some other pine species have leaves with one vascular strand (vein) or five leaves per cluster or sharp-pointed leaves or unarmed cones, only Sugar Pine has all of these characters combined. However, its long cones are probably unique in the genus. The three characters listed at the generic level occur only in *Pinus*, and in all species of pines. The two traits given at the family level are present in the Pinaceae (pine family) but not together in other families. The order (Coniferales), which is the next higher and more inclusive taxonomic rank, is defined (among other things) by the concurrence of the three characters listed there.

10

The genus *Pinus* contains many species; the family Pinaceae has many genera; and the order Coniferales has a few other families in addition to the Pinaceae. In practice, we are generally concerned mostly with the family, genus, and species of a plant.

At a very high taxonomic level, flowering plants are divided into two groups. The Dicotyledoneae, or dicots, are those flowering plants which have flower parts in 4's or 5's, vascular bundles of the stem arranged in a ring-like pattern, and as the name implies, they have two seed leaves or cotyledons. The Monocotyledoneae, or monocots, have flower parts arranged in 3's, scattered vascular bundles, and single seed leaves. There are exceptions to almost all of the above character patterns, but nevertheless the dicots and monocots are distinctive and evolutionarily well-separated groups of plants.

Some familiar dicots are California Poppy (*Eschscholzia californica*), meadow foam (*Limnanthes* spp.*), fiddleneck (*Amsinckia* spp.), lupine (*Lupinus* spp.), and filaree (*Erodium* spp.). Monocots include trillium (*Trillium* spp.) grasses, sedges, yuccas, palms, blue-eyed grass (*Sisyrinchium* spp.), irises (*Iris* spp.), lilies, and various other familiar genera or families.

Naming New Plant Species

The naming of plants is done by taxonomists. Naming of plants is not completely arbitrary, but must follow a series of rules that are laid down in the International Code of Botanical Nomenclature. The Code does not tell a taxonomist what name he must give to a plant or how to determine whether a species is undescribed; it simply gives him the procedure to be followed in naming a plant that he believes is a new species. Although the Code is a fairly lengthy legalistic document, it contains common-sense rules and in general is a practical guideline for taxonomists. Despite the fact that the California flora is rather well known, a number of new plant species are described from the state each year. Some of these new species have been known for many years but have been confused with other closely related species. Intensive studies of the genus

*spp. = plural of species, abbreviated.

11

Clarkia (Onagraceae) — known as godetia or farewell-to-spring — have shown that several "species" that have long been recognized are in fact each composed of more than one species. As a result of these studies, a number of new species have been described in recent years. There are a few species of other genera in California which have been collected for the first time in recent years. Most of these have come from the North Coast Ranges or Klamath mountains, but recently a species of *Limnanthes* (meadow foam, Limnanthaceae) was described from a series of conspicuous populations just north of San Francisco Bay and in a fairly densely populated area. Even more remarkable, a distinctive, previously uncollected new species of the mariposa-lily genus *Calochortus* (*C. tiburonensis,* Plate 2B) was recently discovered on a hilltop on Tiburon peninsula, Marin County, where the small population of the plant grows in full sight of the millions of residents of the San Francisco Bay area! No one had collected specimens of this plant until very recently, although there is reason to believe the species has occurred in this heavily populated region for thousands of years.

In many manuals that cover the California flora, the binomial may be followed by a surname or an abbreviation of the surname of the person who named the species. Examples are *Astragalus nutans* Jones (for Marcus E. Jones) and *Angelica tomentosa* Wats. (for Sereno Watson). Although these surnames are technically a part of the plant name, they are not a part of the binomial and are not used in verbal communication, although in botanical writing they are often used.

2. THE CALIFORNIA FLORA

Size of the Flora

According to Munz' *A California Flora* (1959) and the *Supplement to the California Flora* (1968), there are just over 5,000 vascular plant species in California that are native to the state. (A *vascular plant* is one which has a well-developed vascular system consisting of specialized cells that function to transport water, dissolved minerals, and other substances throughout the plant body. Fungi, algae, mosses, and liverworts are not vascular plants; club mosses, horsetails, ferns, gymnosperms, and flowering plants are vascular plants.) In addition, there are an estimated 975 additional species that have been introduced into the state and have become established as a part of the weed flora of the state. The species listed by Munz are those native and introduced vascular plant species that grow in California without cultivation. In total, the coverage by Munz includes about 6,000 native and introduced vascular plants that are distributed in 1,139 genera. Of the 162 plant families native in the state the following are the six largest (figures are approximate only):

FAMILIES	GENERA	SPECIES
Compositae (sunflower family)	150	850
Gramineae (grass family)	60	465
Leguminosae (pea family)	23	380
Scrophulariaceae (figwort family)	21	315
Cruciferae (mustard family)	35	235
Cyperaceae (sedge family)	12	210
Totals: 6 families	301 genera	2455 species

A simple analysis of the preceding figures reveals that 41 percent of the species of vascular plants in California belong to only six families. Anyone who is faced with identifying California plants in any part of the state will simplify his task by learning the distinguishing characteristics of these six families.

Compositae

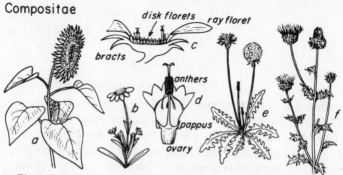

Fig. 1. Compositae. a, sunflower. b, goldfield. c, portion of flower head. d, details of disk floret. e, dandelion. f, thistle.

Gramineae

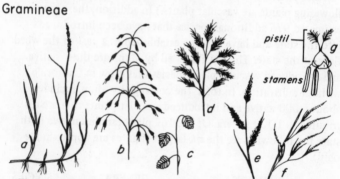

Fig 2. Gramineae. a, rye grass. b, flower stalk of oats. c, three spikelets of rattlesnake grass. d, spikelet clusters of brome. e, grama grass. f, needle grass. g, details of flower.

Leguminosae

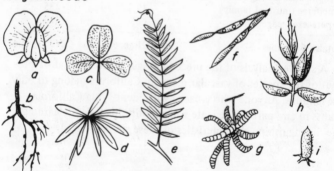

Fig. 3. Leguminosae. a, typical flower. b, roots with bacterial nodules. c, clover leaf. d, lupine leaf. e, vetch leaf. f, open vetch pod. g, pods of mesquite. h, pods of locoweed. i, pod of lupine.

Scrophulariaceae

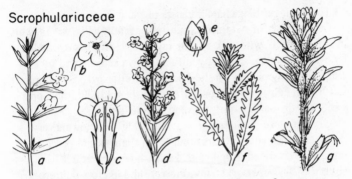

Fig. 4. Scrophulariaceae. a, monkey flower. b, flower of penstemon. c, detail of penstemon flower. d, flowering stalk of penstemon. e, open penstemon capsule. f, lousewort. g, Indian paintbrush.

Cruciferae

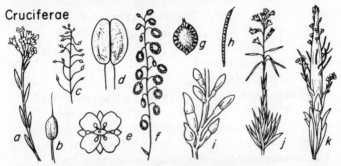

Fig. 5. Cruciferae. a, flowering stalk of bladder pod. b, fruit of bladder pod. c, fruiting stalk of bladder pod. d, peppergrass fruit. e, typical flower. f, fruiting stalk of fringe pod. g, fringe pod fruit. h, rock cress fruit. i, fruits of sea rocket. j, rock cress. k, desert candle.

Cyperaceae

Fig. 6. Cyperaceae. a, flower stalk of sedge. b, stem cross section of sedge. c, detail of typical pistillate flower. d, flower scale. e, detail of staminate flower. f, common tule. g, flower clusters of common tule. h, spike rush. i, umbrella sedge.

The distinguishing characteristics of these families are:

Compositae (or Asteraceae; sunflower family)(Fig. 1): Flowers in a dense head, with disk and/or ray florets. Heads surrounded by bracts. Stamens united by anthers. Calyx represented by a scaly or bristly pappus on the 1-seeded inferior ovary. Examples: sunflower, daisy, aster, ragweed, sagebrush, goldfield, pineapple weed, thistle, balsam root, tarweed, dandelion.

Gramineae (or Poaceae; grass family)(Fig. 2): Flowers very small, greenish, and inconspicuous, with stamens and pistil or unisexual, clustered in spikelets. Perianth greatly reduced or absent. Ovary superior. Stem hollow, round in cross-section; leaves 2-ranked. Examples: cheat grass, brome, rye grass, pampas grass.

Leguminosae (or Fabaceae; pea family)(Fig. 3): Flowers with stamens and pistil, generally bilaterally symmetrical. Stamens usually 10, separate or fused. Ovary superior. Leaves alternate, usually divided into 3 or more leaflets. Roots with bacterial nodules. Fruit a pod. Examples: palo verde, lupine, clover, locoweed, vetch, redbud, mesquite.

Scrophulariaceae (figwort family)(Fig. 4): Leaves undivided into leaflets, usually opposite. Flowers often showy, with stamens and pistils, weakly to strongly bilaterally symmetrical. Ovary superior, fruit a capsule. Examples: *Penstemon*, monkey flower, indian paintbrush, chinese houses, elephants' heads, owl's clover.

Cruciferae (or Brassicaceae; mustard family)(Fig. 5): Leaves alternate, undivided into leaflets although often deeply lobed. Flowers with stamens and pistils, typically with 4 petals and sepals. Stamens 6. Ovary with 2 chambers. Examples: mustard, wall flower, peppergrass, desert candle, sea rocket, water cress, shepherd's purse.

Cyperaceae (sedge family)(Fig. 6): Grasslike herbs of damp places. Flowers with stamens and pistil or unisexual, inconspicuous, clustered in spikelets. Perianth represented by scales or bristles. Ovary superior, often enclosed in a sac. Stem usually solid, triangular in cross section. Leaves usually 3-ranked. Examples: tule, sedge, umbrella sedge.

If you remember the simple characteristics of these six families, plant identification will be greatly simplified, for almost half the plants that you pick up in the field will belong to these families. If you recognize them on sight a considerable amount of time in identifying the plants will be saved.

About 10 percent of the vascular plant species in California belong to six genera: *Carex* (sedges, Cyperaceae); *Astragalus* (locoweeds, Leguminosae); *Phacelia* (most without common names, Hydrophyllaceae); *Lupinus* (lupines, Leguminosae); *Eriogonum* (wild buckwheat, etc., Polygonaceae); and *Mimulus* (monkey flowers, Scrophulariaceae).

Endemism and the California Floristic Province

The term *California Floristic Province* refers to the geographical area which contains assemblages of plant species that are more or less characteristic of California and that are best developed in the state. This province includes southwestern Oregon and northern Baja California but excludes certain of the southeastern California desert regions as well as the area of the state that is east of the Cascade-Sierra axis (see Map 1). The flora of these transmontane or desert areas is best developed outside the state, and therefore, parts of the state of California are not in the California Floristic Province. The Great Basin Floristic Province includes some of the area east of the Sierra Nevada and includes some regions in the northeastern part of the state, although some botanists consider that the latter area belongs to a distinct floristic province which is called the Columbia Plateau Floristic Province.

One striking feature of the flora of the state of California is the high percentage of endemism. An *endemic* species is a plant species which is restricted to a specific locality or habitat. For example, Monterey Cypress is endemic to the Monterey peninsula. *Limnanthes vinculans* is endemic to Sonoma County. Coast Redwood is endemic to the California Floristic Province but not to the state of California, since it also occurs in extreme southwestern Oregon. *Senecio clevelandii* is endemic to serpentine soils.

About 3.4 percent of the genera of vascular plants in California are endemic to the state; about 30 percent of the species are

17

endemic to the state. The genera with the largest number of endemic species are *Mimulus* (monkey flowers), *Astragalus* (locoweeds), *Lupinus* (lupines), *Eriogonum* (wild buckwheats), *Arctostaphylos* (manzanitas), and *Ceanothus* (wild lilacs, etc.). Most of the genera on this list are also on the list of the largest genera in the state.

Some endemics in the California Floristic Province are widespread and are rather well known. Perhaps the most famous endemic of the province is Coast Redwood, *Sequoia sempervirens*. Because of the measures that have been taken to preserve this tree, the Coast Redwood has been saved from extinction, large tracts of this species having been set aside for perpetuity. However, a number of other less spectacular endemics are faced with extinction and some plant species have already become extinct due to man's activity. In the San Francisco Bay region, for example, there is an endemic species of the very small composite genus *Blennosperma*. This is *Blennosperma bakeri*, a species whose entire range is within a small area of Sonoma County. *Blennosperma bakeri* is an attractive small annual related to the more widespread *B. nanum*; it was first described in 1941. At the time of writing, the range of *B. bakeri* consists of one small population surrounded by houses inside the limits of the city of Sonoma plus a second, somewhat larger population at the edge of town. It is almost certain that within the next few years this species will become extinct in the wild, since its habitat will become suburbanized.

Another endemic in the San Francisco Bay area is a member of the fiddleneck genus *Amsinckia* (Boraginaceae). Some species of this genus are widespread and somewhat weedy, but the attractive, large-flowered *A. grandiflora* is now known to occur in only a single, rather small population that is located within the fences of the Lawrence Berkeley Laboratory installation near Livermore. Also in northern California, the attractive *Fremontodendron californicum* subsp. *decumbens* is known only from two populations of a few plants each at Pine Hill, Eldorado County (Plate 1B). *Iris munzii* (Plate 1C) occurs along the Tule River, Tulare County. In southern California, a relative of the Mountain Mahogany is known only as six or seven survivors on Santa Catalina island. This small tree is *Cercocarpus traskiae* (Rosaceae); fortunately, it thrives in cultivation and even if it does become extinct in the wild, it will persist in botanical gardens. Another southern California endemic, one that occurs in the Mojave Desert, is the diminutive annual poppy, *Canbya candida* (Plate 2A). More widespread endemics are California Buckeye (*Aesculus californica*, Plate 2D) and Digger Pine (*Pinus sabiniana*, Plate 2C).

The largest number of plant species endemic to the state is found in southern California, followed by the central coastal area in second place. Relatively few endemics occur in the desert areas, the northern Sierra-Cascades, and the Central Valley.

The reasons for the narrow geographical ranges of California endemics are unclear. Some endemics, such as the Ione Wild Buckwheat (*Eriogonum apricum*; Polygonaceae), occur only on specialized soil types which have narrow distributions. Other species, such as Coast Redwood, are limited by climatic factors – in its case the cool fog zone of the coast. But in most cases, the reason for the rarity is not clear. It is probable that a number of California endemics are evolutionarily "old" species that were more widespread in the geological past and are now on a natural road to extinction (and are often speeded along this road by man). We know, for example, that *Cercocarpus traskiae* of Santa Catalina Island has a fossil history going back several million years. It is likely that this species (like a number of island endemics) has been unable to adjust to the climate of modern California and that this is the reason it is so rare and unsuccessful. Likewise, Coast Redwood and Sierra Big Tree (*Sequoiadendron giganteum*) have extensive fossil records indicating that in past millenia both were once more widespread than they are now. Brewer Spruce (*Picea breweriana*), Santa Lucia Fir (*Abies bracteata*), and Catalina Ironwood (*Lyonothamnus floribundus*) also all have extensive fossil records indicating that these are "old" species that were once much more widely distributed.

At the other extreme, a number of rare plant species are undoubtedly newcomers in an evolutionary sense. For example, many of the newly described, highly restricted species of *Clarkia* from the southern Sierra Nevada foothills are recently evolved. This is also true of several species of tarweeds (*Madia, Hemizonia,*

19

both Compositae) in California. Their restricted distribution is probably a result of the newness of these species. They have not yet developed the genetic variability to allow them to expand their ranges, and in many cases it is unlikely that they will expand their ranges because of the unavailability of suitable habitats.

A number of California endemic plant species are being preserved in state parks or other areas. In addition, native plant gardens at Rancho Santa Ana (Claremont), Santa Barbara, University of California (Berkeley), and Tilden Park (Berkeley) also contain a number of rare and endangered species. Although it is relatively easy to maintain perennial plants in these gardens, maintaining stocks of annuals presents difficulties, since these must be grown from seed each year. The California Native Plant Society, headquartered in Berkeley, is an organization actively involved in the preservation of the native flora of the state.

Flora and Vegetation

The term *vegetation* refers to the life-forms or general aspect of the plants of an area. On the other hand, information about the *flora* of an area (that is, the kinds of plants there) tells you what plant species occur there, but does not communicate much information concerning the "looks" of the plant cover of an area, i.e., whether the area is densely forested, a grassland, or a parklike woodland. For example, the flora of California's Kern County consist of slightly over 1700 plant species, but the list of these species does not give much of an idea concerning the nature of the plant cover of the county. The vegetation of Kern County, however, consists of a variety of different types such as grassland, pine forest, and freshwater marshes, and these descriptive terms provide more information as to what the dominant plants of the area are (e.g., grasses vs. pine trees). However, note that "pine forest" as a term can refer to forests dominated by pine trees over much of the northern hemisphere: the kinds of pine trees at Point Reyes (Bishop Pines), for example, are different from those of the White Mountains (Bristlecone Pines). Thus, with respect to pines, the floras of these two areas are different even though their vegetation type might be considered the same.

The vegetation type and the aggregations of plant species found in an area are determined by a combination of factors:

1. Biotic factors: includes the presence and effects of man, other animals, and other plants.
2. Soils: physical and chemical composition.
3. Physiography: slope, terrain, etc.
4. Climate: temperature, radiation, wind, water, fire.

Biotic Influences on Vegetation

In the past migrations and distributions of various major genera and families, certain continents or large geographical areas were "left out". In other areas (such as the Arctic) the stringent climate prevents certain life forms (such as tall trees) from being present, so obviously no forest vegetation can develop.

An important influence on the composition of vegetation is the effects plants have on each other and the effects animals or micro-organisms have on plants.

In the redwood forest, as well as other coniferous forest types, the intense shade cast by the large evergreen trees undoubtedly is a prime influence in preventing the establishment of a number of plant species under these trees. Even the light shade cast by a deciduous tree standing in a field may have an important effect on determining what plants can grow under the tree.

A second effect that plants have on each other is expressed via competition for water. Sometimes this is on a "first come, first served" basis, that is, the oldest individuals in a community may have well-established and pervasive root systems that prevent the establishment of either their own seedlings or seedlings of other species because these systems are so efficient in taking up water. This effect is particularly important in arid regions where water supplies are generally at a premium. It is probable that the wide, orchardlike spacing of many desert shrubs is a by-product of competition for water.

The presence and activities of fungi and bacteria may also have an effect on the nature of the vegetation. In warm, wet regions decomposition of leaves that fall from trees is very rapid; in cooler or drier areas this decomposition — which is the result of bacterial and fungal action — may be slowed down so considerably that a thick layer of "duff" accumulates under trees. The presence of duff may prevent the establishment of seedlings simply because it is so loose and well aerated that the roots of newly germinated seedlings dry out before they reach the soil surface. In other instances, however, the presence of humus due to sluggish decomposition favors the growth of some plants, expecially saprophytic orchids and relatives of the heather family. (A *saprophyte* is a plant which obtains much of its nutritional requirements from decomposing organic material, such as dead leaves or wood.)

Another plant-plant interaction is one which has received much attention in recent years. This is the phenomenon called *allelopathy*, which can be defined fairly accurately as chemical warfare among plants. Recent work on various native and introduced California plants indicates that certain species give off chemical compounds that are toxic to other plants (and in some instances are toxic also to seedlings of the same species). An example of a genus which shows allelopathic behavior is *Salvia*, a member of the mint family which is commonly known as sage (not sagebrush). Shrubs of *Salvia* produce volatile odorous chemicals called terpenes. These terpenes are given off into the air and also diffuse or wash into the soil and inhibit the establishment of seedlings in the immediate vicinity of *Salvia* shrubs. As a result, stands of *Salvia* contain individuals that are rather uniformly spaced, much as if they had been planted by man. The advantage of allelopathic inhibition of other plants is probably that by preventing other plants from growing in their vicinity, the *Salvia* shrubs insure themselves access to the local groundwater supply. Another plant that exhibits allelopathy is Chamise (*Adenostoma fasciculatum*), a shrubby member of the rose family that is an important constituent of Chaparral in California. There are other examples from the California flora which suggest that allelopathy may be an important factor in influencing the presence and distribution of various plant species in California plant communities.

Another biotic factor is the presence and activities of man and other animals. Without doubt, one of the most important animal effects on the California flora has been the direct or indirect activities of man. Clearing of the land for agricultural purposes, roads, or housing has resulted in the destruction or alteration of large areas of California. Man's accidental (or intentional) introduction of various weedy plant species also has had an important effect on native plant communities. His introduction of grazing animals such as sheep, goats, and cattle, as well as his indirect effect on increasing population size of native herbivores such as deer and rabbits by destroying predators, have had an important influence (generally a negative one) on California plant communities.

There are some plant communities that have been all but eliminated in California as a consequence of man's activities. At one time large tracts of the Central Valley were occupied by a grassland composed of various genera of perennial native grasses. With the advent of agriculture and particularly of grazing domesticated animals, the native grasses disappeared and have largely been replaced by introduced annual grasses as well as by a few introduced non-native herbs (such as filaree, a name applied to various *Erodium* spp.). These introduced grasses are satisfactory for forage purposes, but they are a rather poor substitute for the natural and more enduring plant community that they have replaced (Plate 9D). Perhaps one slightly favorable by-product of man's interference with the natural ecology of large areas of the state is that the display of spring wildflowers may have been locally enhanced by the eradication of native perennial grasses. There is evidence that the abundance of various colorful native herbs is greater in some man-made grasslands than in natural native grasslands. Probably in the latter it is difficult for seedlings of annual herbs to become established because of the extensive root systems of perennial grasses, whereas in the introduced grasslands, the grasses, like the herbs, are annual and may offer less competition for the native herbs because all start growing at approximately the same time.

Other influences of man on vegetation are the pollutants released by industry and automobiles that are resulting in the death of Ponderosa Pine trees over extensive areas of southern California.

Soils and Vegetation

The properties of soils are primarily influenced by their mineral content, the amount of organic matter present, abundance of water, and degree of aeration. Some other features of soils that reflect the presence and nature of these properties are the soil texture, the proportion of mineral nutrients present, the availability of nutrients, heat-exchange capacity, presence of salts, level of the water table, moisture-retaining capacity, presence of micro-organisms, and so on. The complex interactions among these various factors influence the overall nature of the soil and this in turn, has an important effect on the plant cover present in that soil type.

In addition to the factors listed above, the nature of the parent rock from which the soil has been derived is also important. In California, there are large expanses of granitic rock, basalts, serpentine, and even limestone. The nature of the parent material has a strong influence on the type of soil that is ultimately derived from these rocks. Climate, too, has an important effect in determining soil types. In very wet climates, the soil that results from the weathering of limestone may be quite different from that which develops in an arid climate. Topography is also important; deep soils may never form in steep areas.

One soil property that has an important effect on plants is degree of salinity. In many areas of the deserts of California and even in the Central Valley, there are tracts of land that are sometimes called *alkali flats*. In such areas, the soil contains high amounts of various salts of potassium and/or sodium, as well as of other soluble minerals. When these soils are dry, the minerals ("alkali") may leave a whitish or greyish crust on the soil surface. In small depressions or other low areas where water collects during the rainy seasons, the salt concentration may become so high that growth of most flowering plants is inhibited. The higher the salt concentration, the fewer the plant species that can tolerate the soil. After a certain concentration has been reached, no vascular plants can grow, resulting in patches of soil devoid of vegetation. These barren areas are sometimes called "alkali scalds" (Plate 3D). In the Central Valley they may be only a few

24

square feet in extent, but in drier desert regions they may be several acres or even square miles in area.

Although limestone is not as abundant in California as it is in many other states of the United States, there are extensive outcrops of this rock in various parts of the state and certain peculiarities in plant distribution are associated with these outcrops. Although the rocks of the central Sierra Nevada are largely rather sterile granitics, in the Convict Creek basin area between Lee Vining and Bishop, Mono County, there are some marble deposits located between 8,000 and 12,000 feet (2,438 and 3,658 m) in elevation. These alkaline rocks are mostly not forested since there is rather little soil associated with them. However, in the creek basin there are several plant species occurring on limestone that are slightly to extremely out of their normal geographical range. Among these are Bear Berry, *Arctostaphylos uva-ursi*, a low shrub of the heather family (Ericaceae), which occurs otherwise in California only along the coast of the state north of San Francisco Bay and is not known to occur elsewhere in the Sierra Nevada; *Kobresia myosuroides*, a small sedge (Cyperaceae) whose nearest populations are in the Wallowa Mountains of northeastern Oregon and the Uinta Mountains of Utah, and *Scirpus pumilus*, another small sedge whose other populations occur about 750 miles away in the mountains of Colorado and Montana. The explanation for these unusual plant distributions associated with the Convict Creek basin marble deposit is not clear, but it is apparent that in some way the peculiar local soil allows these plants to grow in an area where they otherwise would not be expected.

In the arid regions of California, subtle differences in soil characteristics may have striking effects on the distribution and nature of vegetation types and of plant communities. The various soil types found in California are associated with vegetational effects that range from the undetectable to the very dramatic. To illustrate the effect that soil factors may have in influencing plant distribution, the vegetation of serpentine soils will be discussed in some detail.

The term *serpentine* generally is applied in California to a class of rocks that are essentially magnesium silicate. Serpentine

25

rocks tend to be greenish or gray-green, rather glossy, and to occur in outcrops that range from a few square feet to many square miles in extent (Plate 3B). Serpentine outcrops are common in the North Coast Ranges of California, particularly in Lake County, and reappear in many areas around the Bay Region such as the Berkeley and Oakland hills, Mount Tamalpais, and even the Presidio in San Francisco. South of the Bay, serpentine is perhaps less common than northward, although extensive outcrops occur near San Jose, at Pacheco Pass, and in the region of Idria in San Benito County, and in various other portions of the South Coast Ranges. Serpentine also occurs in various portions of the foothills of the Sierra Nevada, and can be seen in many areas near and along Highway 49. Although the raw outcrops are generally greenish in hue, weathered serpentine and the soil derived from it often are red. Many of the mountains in the state named Red Mountain bear reddish soil derived from serpentine. Serpentine is sufficiently distinctive and widespread in California that in 1965 the legislature designated it as the official state rock.

The discontinuity between serpentine vegetation and the vegetation on adjacent non-serpentine soils generally is very striking (Plate 3C). Where the two soil types meet, dense forest may give way abruptly to open chaparral, or even to large expanses of ground that support a few shrubs interspersed with occasional herbs. In the North Coast Ranges, serpentine often supports chaparral in a climatic region that, on other soils, favors the development of coniferous forest.

Serpentine soils are unproductive from an agricultural standpoint. Few crops can be grown successfully, nor do the soils support the growth of forage grasses for grazing animals. Since trees usually occur only rather sporadically and thinly on serpentine, such areas also are not useful for timber purposes.

Characteristics of serpentine soils in California that are unfavorable to plant growth are:

1. Serpentine soils have a very low calcium content and a correspondingly high magnesium content. Calcium and magnesium are both essential nutrients in the metabolism of plants, but plants are unable to take up sufficient calcium through their

roots when magnesium is present in excessive quantities. The result of this imbalance may be stunted growth of plants due to a combination of calcium deficiency and magnesium toxicity. By addition of calcium salts, serpentine soils may be rendered relatively amenable to agricultural use.

2. Serpentine soils also frequently have a high nickel and chromium content. These mineral elements are not only unnecessary for plant growth, but are toxic to plants even in small quantities.

3. Serpentine soils are low in the content of nutrients, such as nitrogen, that are required in relatively large quantities by most plants for adequate growth. In addition, they are deficient in many other nutrients, such as molybdenum, which are equally essential to plants but are needed in smaller quantities.

4. Serpentine soils often are waterlogged in the winter and excessively dry in the summer season. The transitional periods between being very wet and being very dry may be quite short, with the result that a perennial plant growing on serpentine soil must be able to tolerate very wet soil during some seasons of the year and very dry soil during other seasons. Such extremes in water content also may be characteristic of other soil types in California, but few other soils offer as well the peculiar set of characteristics of mineral composition.

Serpentine soils have a very high proportion of endemic plant species restricted to them. For example, Sargent Cypress (*Cupressus sargentii*, Cupressaceae) is found only on serpentine deposits in the Coast Ranges from Mendocino County southward to Santa Barbara County. It is therefore termed an obligate serpentinophile. Many species of jewel flower (*Streptanthus* spp., Cruciferae) likewise occur only on serpentine, and some of these species are known only from a single serpentine outcrop. These species, too, are obligate serpentine endemics. There are all gradations of fidelity to serpentine, ranging from 100 percent to nil. Some species, such as Leather Oak (*Quercus durata*; Fagaceae) of the North Coast Ranges, commonly are found on serpentine, but populations also occur on volcanic soils in Napa

27

and Sonoma counties. Similarly, Macnab Cypress (*Cupressus mac-nabiana*, Cupressaceae) is also generally, though not always, found on serpentine. Clearly, neither of these tree species is a serpentine endemic, although both of them usually occur on serpentine soil. Next in line are species such as Knobcone Pine (*Pinus attenuata*, Pinaceae) which frequently grow on serpentine, but commonly occur on other soils as well. The fidelity of this pine to serpentine is perhaps only about 50 percent. Next is a species such as *Jepsonia heterandra*,(Saxifragaceae) whose populations often occur immediately adjacent to serpentine rocks, but rarely on serpentine. Only a few, rather unhealthy individuals of this species are established on serpentine soil even though abundant seed is dispersed to this soil type. Finally, there are many plant species in California which never occur on serpentine. Therefore, some species in the state's flora are true serpentine endemics; others may be found on serpentine in some areas; and still others are not known to occur on serpentine anywhere in their range.

Indicator Species

The idea that certain plant species may be restricted exclusively to a single soil type or, by extrapolation, may occur in only a certain climatic regime (such as does Coast Redwood, *Sequoia sempervirens*) brings up the concept of an indicator species. The idea goes back to the American ecologist F. E. Clements, who dominated the American ecological scene for many years. Although many of Clements' ideas have fallen into disrepute in recent years, he was an original thinker and made a number of useful contributions to the development of ecological theory. In Clements' words, the idea of an *indicator species* grows out of the statement that "Every plant is a product of the conditions under which it grows and is, therefore, a measure of environment." One can, as a consequence, state that a species such as Sargent Cypress is an indicator of serpentine soil. The goldfield *Lasthenia minor* subsp. *maritima* (Compositae) is an indicator of seabird guano (Plate 4C). Cattails (*Typha* spp., Typhaceae) are an indicator of prolonged (or permanent) flooding. The

Greasewood of the Great Basin (*Sarcobatus vermiculatus*, Chenopodiaceae) is an indicator of saline soils. In many areas of the West, various species of brome grass (*Bromus* spp., Gramineae) are indicators of overgrazing, as is Basin Sagebrush (*Artemisia tridentata*, Compositae). Knobcone Pine is an indicator of fire within recent decades. Thus, the presence of certain species in an area provides insights into the ecological status and history of the area. The same is true of a fire-induced plant community; its presence in areas that would otherwise support other plant communities is an indication of fire in the history of the area. The presence of Coast Redwood is an indicator of extensive coastal fogs. The Alkali Sink Scrub with its characteristic assemblage of desert shrubs is, as its name implies, an indicator of alkaline, or at least, saline soils. The presence of certain indicator species tells a great deal about the general ecological characteristics of an area; the concept is particularly important to agriculturalists who may be interested either in the future agricultural prospects of an area or in whether a plot of land is being properly managed.

Ecological Races

Some years ago, A. R. Kruckeberg, then a graduate student in the Department of Botany of the University of California, Berkeley, devised a series of experiments aimed at testing the presence of ecological races within certain plant species of California that were known to occur both on serpentine (in some areas) and off serpentine (in other areas). Kruckeberg filled several planter boxes with serpentine soil and others with ordinary garden soils. Among the plants Kruckeberg worked with was an annual blue- or white-flowered gilia, *Gilia capitata* (Polemoniaceae, Plate 3A). This species has populations that occur on serpentine and others that occur on other soil types; plants from the two soil types are indistinguishable morphologically. Kruckeberg collected seed from serpentine and from non-serpentine populations and grew the offspring on both serpentine and non-serpentine soils. In general, seedlings from non-serpentine populations grew very well on non-serpentine soils, but did very poorly on serpentine soils. The latter either died shortly after gemination or grew very slowly. A few managed to make it through to flowering, but only as dwarfed plants. Seedlings from serpentine populations grew very well on serpentine soils and also did very well on non-serpentine soils. One conclusion that was drawn from this set of simple experiments is that *G. capitata* is made up of at least two genetically different ecological races adapted to different soils. One of these races is a serpentine race which tolerates serpentine soil (as well as non-serpentine soil) and the other is a serpentine-intolerant race which can grow only on non-serpentine soils.

One question that might be asked, however, is that if the serpentine races can grow not only on serpentine soils but off serpentine soils as well, why is it that apparently all plants of *Gilia capitata* that occur off serpentine in the wild are, in fact, serpentine intolerant? The answer to this seems to be that, for some reason, the serpentine races are not only tolerant of serpentine soils but they are generally restricted to these soils because they are susceptible to pathogenic soil fungi that occur in non-serpentine soil. Whenever seeds of the serpentine race stray over to non-serpentine soils the resultant seedlings die because they are attacked by pathogenic soil fungi present under field conditions, or are otherwise unable to compete successfully with plants on these soils. In contrast, the serpentine-intolerant races of *G. capitata* are not only intolerant of serpentine but are tolerant of the soil fungi. One might characterize *G. capitata* as a species made up of serpentine-tolerant, fungus-intolerant races and serpentine-intolerant, fungus-tolerant races. R. B. Walker, a contemporary of Kruckeberg, suggested that the physiological basis of serpentine tolerance lay chiefly in the ability of serpentine tolerant races to extract sufficient calcium for their metabolic needs against the magnesium gradient.

Regardless of the physiological interpretations that can be made from the series of experiments and observations described above, it is clear that soils do have an important influence on plant distribution, and it can be demonstrated clearly that some species or races of plants can survive only on specific soil types and are genetically and physiologically adapted to them.

Another example of the effect that a peculiar soil type may have on the distribution of a plant species is seen in the distribution of Ione Manzanita, *Arctostaphylos myrtifolia* (Ericaceae, Plate 1D). This small shrub is restricted to the Ione formation, which is a highly acid soil that is a mixture of clay, sand, and ironstone. The Ione formation occurs in a small area in the vicinity of Ione, Amador County, about 40 miles (64.4 km) southeast of Sacramento. Another endemic of the Ione formation is the Ione Wild Buckwheat, *Eriogonum apricum* (Polygonaceae). The ecological reasons behind the restriction of these two plant species to the Ione formation are not clear, but they nevertheless provide another example of endemism that is related to soil type.

A rather peculiar example of endemism to another soil condition comes from the goldfield genus *Lasthenia* (Compositae). Many species of this genus form large and colorful populations in the valley regions of mainland California. However, *L. minor* subsp. *maritima* is restricted to the guano-rich soils of offshore islands, ranging from the Farallon Islands west of San Francisco northward to some islets at the northern tip of Vancouver Island (Plate 4C). So far as is known, this subspecies grows only on nitrate-rich seabird guano. Tests of its foliage indicate that this plant is a nitrate-accumulator, which suggests that it is able to tolerate the high concentration of nitrates in soils that are toxic to most other plants. It is probable that the distribution of this subspecies is aided by the migration or casual flights of sea birds from island to island along the Pacific coast. In cultivation, *L. minor* subsp. *maritima* grows well on normal garden soil, but in the field it seems unable to compete with other plant species on any substrate but the highly friable and malodorous seabird guano.

The serpentine investigations of Kruckeberg, Walker, and others, plus the investigations of a number of other biologists and soil scientists, suggest the following general conclusions concerning plants and soils:

1. Where soil patterns are complex, the vegetation patterns are correspondingly complex.

2. The presence of certain atypical soils (such as serpentine, acid clays, etc.) may result in the presence of a plant community that is not typical of the general area.

3. The restriction of a plant species or a race to a specific soil type may be due to genetically determined physiological adaptation to this soil as well as to an inability of the plant to survive naturally on other soil types.

4. The absence of a plant species (or race) from a soil type present within its geographical range is probably due to a genetically determined intolerance to this soil type.

5. A plant species that occurs on a wide variety of soils can do so because it is made up of ecological races that are individually adapted to these soils. Each ecological race may have a narrow adaptation, and one result of this is that the ecological amplitude exhibited by a species is a rough reflection of its ecotypic richness. That is, a species that occupies a wide range of habitats is likely composed of more ecological races than is a species which occurs in a more restricted variety of habitats. Thus, in general, the ability of a species to occupy a wide range of habitats is due to the presence in the species of ecological races, otherwise called *ecotypes*.

There are different kinds of ecotypes. For example, *Gilia capitata* contains at least two different soil ecotypes, one adapted to serpentine soils, the other to non-serpentine soils. *Lasthenia minor* likewise has two soil ecotypes: one of these (sufficiently distinct in its external morphology that it is recognized as a taxonomic subspecies) occurs only on guano (Plate 4C) and the other (another subspecies) is on non-guano soils. There are also climatic ecotypes, that is, races of species adapted to different climates within the range of the species. Many wide-ranging tree species are made up of climatic ecotypes, as are several herbaceous species (such as Yarrow) that occur from sea level to above timberline in the Sierra Nevada. There are also seasonal ecotypes: in the California tarweed *Madia elegans* there is one spring-flowering ecotype and another fall-flowering ecotype.

The term ecotype is a useful one, but must be used with three qualifications in mind. The first of these is that "ecotype" must be used with some sort of qualifier indicating what ecological condition is involved. One must speak of seasonal, climatic, soil, or other ecotypes. Second, an ecotype is something that can be identified initially only by experimental methods. One cannot look at a plant in a specific habitat and term it an ecotype in the absence of any information concerning the plant. Third, the term ecotype is used only at the infraspecific level. One cannot speak of Ione Manzanita (*Arctostaphylos myrtifolia*) as an ecotype adapted to Ione clay, since the term ecotype is used only when two or more ecological races occur *within* a species. Also, it would be incorrect to call *Gilia capitata* an ecotype. *Gilia capitata* contains at least two soil ecotypes, one of which is a serpentine ecotype and the other of which is a non-serpentine ecotype. There may be climatic ecotypes in *G. capitata,* too, although these have not yet been demonstrated.

The preceding discussion has been concerned largely with soil effects on vegetation and plant communities, although more emphasis has been placed on individual species than on the communities of which they are members. The third and fourth factors that influence the presence and nature of plant communities and vegetation are climate and physiography. Since climate is one of the prime determinants of plant distribution and since physiography is an important determinant of regional climate, the two factors will be discussed together rather than as independent entities.

3. TOPOGRAPHY, CLIMATE, AND ADAPTATION

Topographical Features of California

The area of California is about 160,000 square miles (ca. 414,400 sq. km), most of which land supports plant life. The state is approximately 800 miles (ca.1,287 km) long and about 200 miles (ca.322 km) wide. Elevations range from about 280 feet (85 m) below sea level in Death Valley to approximately 14,500 feet (4,420 m) on the top of Mount Whitney, not far from Death Valley. The area is topographically diverse, but has as its major feature two mountain systems extending in a north-south direction, the Coast Ranges on the west side of the state and the Sierra Nevada-Cascade ranges on the east (Map 1). The two systems join in the north in the vicinity of Mount Shasta and also in the south at the southern end of the San Joaquin Valley. The relatively flat valley delimited by this ring of mountains, the Central Valley, is made up of the Sacramento Valley in the north and the San Joaquin Valley in the south. The two great rivers of these valleys, the Sacramento and San Joaquin rivers, join east of San Francisco Bay and flow westward through the Bay into the Pacific Ocean.

The Coast Ranges consist of the North Coast Ranges north of San Francisco Bay and the South Coast Ranges south of the bay. At the northern end of the state, the Coast Ranges merge into the Klamath mountain massif which in turn merges with the southern end of the Cascades. This complex of mountains forms an important climatic and phytogeographical barrier at the northern end of the Sacramento Valley. In the southern portion of the state, the South Coast Ranges are joined with the southern Sierra Nevada via the Transverse Ranges, which include the Tehachapi Mountains. In extreme southern California there are the Peninsular Ranges, which extend across the Mexican boundary well into Baja California.

The important aspects of the chief topographic features of the state are:

1. *The Coast Ranges:* In general, the Coast Ranges are a relatively old range of rather low mountains that rise abruptly at the immediate coastline of the Pacific Ocean. The orientation of these ranges is north to south. Between the major north-south ridges there are numerous valleys ranging in size from very small ones to sizable ones such as the Napa, Sonoma, and Salinas valleys. The Coast Ranges are interrupted at San Francisco Bay: the mountains of this system that occur north of the bay are called the North Coast Ranges, those south of the bay are called the South Coast Ranges.

2. *Sierra-Cascade axis:* Geologically these two mountain systems have had different origins, but since they form a continuous north-south high mountain range they can be considered together. The Cascade Range, which is extensively developed in Oregon and Washington, reaches its southern limit in northern California and is terminated there by Mt. Lassen (10,453 ft, 3,186 m elev.). The range is volcanic in origin throughout most of its area. The Sierra Nevada is exclusively a California range which is generally non-volcanic in origin and which is characterized by the occurrence of immense expanses of granitic rocks. These mountains originated by faulting. They rise gradually to a crest that is frequently well above 10,000 feet (3,050 m), and the range drops off very sharply to the east. Unlike the Cascades, the Sierra Nevada range consists of a series of ridges, with the major peaks (such as Mount Whitney) rather poorly differentiated from the surrounding peaks. There is evidence of extensive glaciation in the higher reaches of the Sierra Nevada, and this glaciation is believed to have had an important effect on the present distribution of some woody plants in these mountains.

The Sierra-Cascade ranges are heavily vegetated, with various forest types on the lower slopes and with alpine or other montane herbaceous or shrubby vegetation in the upper regions. Despite the difference in geological origin between the ranges, there is relatively good continuity in the vegetation types running from one range into the other.

3. *Klamath Mountains:* These are a geologically and topographically complex series of mountain ranges that occur in

northwestern California and adjacent southwestern Oregon. The mountains in this area are topographically rather rough, frequently accessible only with difficulty, and floristically very interesting. There is a variety of soils in this area, including serpentine and limestone, and because of this there is a relatively high number of plant species endemic to these moderately high mountains.

4. *Transverse Ranges:* The Transverse Ranges consist of relatively low mountains that run in an approximately east-west direction in the southern one-fourth of the state. They extend from near the Pacific Ocean eastward across the southern end of the San Joaquin Valley and ultimately join with the southern Sierra Nevada.

5. *Peninsular Ranges:* This mountain system in the extreme southwest section of the state is probably to be considered as a southern extension of the South Coast Ranges, although the continuity is obscured by the intrusion of the Transverse Ranges discussed above. These mountains are relatively low, and like the Sierra Nevada, offer extensive areas of granitic soils.

6. *Central Valley:* This wide, flat valley occupies the central portion of the state between the Coast Ranges and the Sierra-Cascade axis. This valley is very flat and in some areas has extensive marshes. Many of these have been drained for agricultural purposes, but remnants still may be seen in various parts of the valley. Before the development of intensive agriculture, the Central Valley contained a number of interesting plant communities, but these are now largely a thing of the past and we must rely on historical records concerning their nature and extent. The soils of the Central Valley are variable and may consist of peat, sands, rich alluvial soils, alkaline clays, or rather sterile hard soils such as those that occur in some of the oil-producing areas of the San Joaquin Valley.

7. *Deserts:* In the southern portion of the state and adjacent regions there are two major deserts:

A. *The Colorado Desert:* This is a low desert east of the Peninsular Ranges that is presently dominated by the Salton Sea. It extends into adjacent Mexico.

B. *The Mojave Desert:* The Mojave Desert is a somewhat higher desert than the Colorado Desert and is broken up by a number of small mountain ranges. This desert extends into adjacent Nevada. It is rather poorly drained, and as a consequence there are numerous periodically inundated low areas that are now highly saline because of the influx of water during very wet years followed by evaporation of their contents in subsequent years.

8. *Great Basin and related areas:* North of the Mojave Desert and east of the Sierra Nevada are occasional valleys or upland areas that are basically the westernmost reaches of the Great Basin, which is a large, relatively high area that lies between the Sierra Nevada (and Cascades) and the Rocky Mountains and has its center in Nevada. To the north, in northeastern California, the Modoc Plateau is an area in which there are extensive areas covered by lava.

Climate of California

The climate of California is strongly influenced by the physiography of the state. Botanically, climate may be defined as the sum total of atmospheric conditions that influence the growth and reproduction of plants. In much of the United States, the latitude approximately fixes the regional climate. While this may be true in a general sense for California, the tremendous variation in topography exerts an influence which is equal to, if not more important than, the latitude. In California, the major factors determining the climate are the Pacific Ocean and the presence of mountainous masses of land.

Much of California has a maritime climate; this is in contrast with the continental climate of montane or inland areas. The following shows the extremes between these two climatic regimes:

MARITIME CLIMATE	CONTINENTAL CLIMATE
Coastal location	Inland location
Winters warm	Winters cold
Summers cool	Summers hot
Small daily temperature range	Large daily temperature range
Small seasonal temperature range	Large seasonal temperature range
High relative humidities	Low relative humidities

The climate of most cismontane California, that is, of that part of California lying west of the Sierra-Cascade crest, is a *Mediterranean* one. The summers are cool and dry; the winters are relatively warm and wet. There are even some genera, such as rock rose (*Helianthemum*, Cistaceae), sage (*Salvia*, Labiatae), and tree mallow (*Lavatera*, Malvaceae) that are common to similar climatic zones of California and Mediterranean lands, though different species occur in each of the two regions.

The climatic influences of the ocean are manifold. Low pressure areas that develop in the Pacific Ocean in the vicinity of the Gulf of Alaska are stationary during some parts of the year, but in the winter they frequently move southeasterly and bring cold weather, strong winds, and rain to much of California. In the summer, these Pacific lows are generally centered to the northward and as a result summer rainfall is rare in California, except for local thunderstorms in the mountains. During the summer, the moisture present in the cold oceanic air often condenses when this air meets the warm, dry air from the land. The result is the development of the coastal fog belt that is typical of much of the California coastline during summer months.

In most of the central and eastern United States, climatic zones tend to follow roughly latitudinal lines running from east to west across the continent. In California, however, the climatic zones generally run in a more or less north-to-south direction. This is due to the strong influence of the Coast Ranges and the Sierra-Cascade axis: since these mountains run in a north-south pattern, so does the climate over which they exert such a strong influence. For example, the winter storms that bring rain to California generally come into the state from the Pacific Ocean. The distribution of this rain within the state, however, is largely determined by topography. Much of the moisture in the Pacific air is dropped by the time it reaches the crest of the Coast Ranges, and a second portion of moisture is lost when the air reaches the crest of the Sierra-Cascade ranges. As a result, one finds that the immediate coast and the westerly slopes of both ranges have a higher precipitation than do areas slightly east of each. Since the Sierra-Cascade ranges are considerably higher than the Coast Ranges, rather little moisture-laden air reaches beyond this mountain axis, and the result is that the Great Basin has a considerably lower winter rainfall than does that portion of California immediately to the west of these mountains.

In the same fashion that these mountains act as a barrier to the passage of moisture-laden winds, they also act as barriers to the passage of hot or cold air masses. During the summer, the Sierra-Cascade ranges protect much of California from the hot, dry air masses that develop over the central United States. This, combined with the proximity to the cool Pacific Ocean, explains why California has a generally cool summer climate. During the winter, these mountains serve to insulate California from the cold, dry air masses that develop over the inland portion of the continent. As a consequence, California has winters that are in general milder than one might expect at the latitude.

In going up the western side of the Sierra Nevada, one encounters an increase in annual (mostly winter) precipitation up to a point at which it begins to drop off as one continues to ascend in elevation. The zonation of vegetation in the mountains reflects this variation in precipitation patterns. In the lowest portions of the foothills the vegetation types are characteristic of a relatively arid climate. In middle elevations the vegetation reflects the rather favorable precipitation patterns. Above the middle elevations, however, the vegetation acquires a more "arid" aspect. As one goes southward in the Sierra, these climatic zones move up in altitude, with the result that in the southernmost Sierra Nevada the zone of highest precipitation is considerably higher than it is in the northern Sierra.

Of chief importance to the distribution and nature of the plant cover of California is the amount and seasonal distribution of rain or other precipitation. The wettest portions of California are in the northwestern part of the state. Wet areas also occur on the western slopes of the Sierra Nevada and in the Santa Lucia Range of Monterey County. On the western slopes of the Coast Ranges, from coastal Monterey County northward to southern Oregon, the average annual rainfall exceeds 50 inches (127 cm) per year. This is also true for much of the western slope of the Sierra-Cascade ranges. On the other hand, the Sacramento Valley averages below 20 inches (51 cm) of rain per year and the San

Joaquin Valley has less than 10 inches (25 cm). The southeastern deserts receive an average of less than 5 inches (13 cm) per year, and in a few areas, such as Death Valley, some years may pass by without any measurable rainfall. The extremes in average annual rainfall range from an excess of 110 inches (280 cm) in parts of Del Norte and Siskiyou Counties to less than 2 inches (5 cm) per year at Furnace Creek Ranch in Death Valley.

Most of the rainfall that occurs in California falls during the winter months. The rainy season in southern California is generally during a period of five months between November and March; in northern California it is during a seven-month period between October and April. There are as many as 100 days of measurable rain on the average in parts of northern California; there are as few as 10 days in some desert regions. In the calendar year 1909, over 153 inches (389 cm) of rainfall were recorded at Monumental in Del Norte County; during one season (July 1 – June 30) over 160 inches (406 cm) of rain were recorded at one station in Monterey County. In southern California, there are frequent intense seasonal storms during which very heavy amounts of rain fall. For example, one storm in Los Angeles County in late January, 1943, dropped nearly 26 inches (66 cm) of rain in 24 hours. On another occasion, over 11 inches (28 cm) of rain were recorded in 80 minutes at Campo, San Diego County. Obviously, such heavy amounts of rain during very short periods are of little value to plants and may result in damaging floods in lowland areas.

Although the average rainfall for much of the state is so low that one might expect few plants to survive such conditions (because it falls mostly during the winter months), many arid areas provide shows of spectacular annuals which flower in early spring. Such displays are conspicuous in the southern end of the San Joaquin Valley and also in portions of the Mojave and Colorado deserts. The average rainfall figures for a particular area may not be very helpful in estimating what sort of a vegetation is present in the region, however, because of the yearly fluctuations in rainfall and because of the seasonal distribution of the rain. Prolonged periods of drought may have a negative effect on the survival of woody plants, with the result that shrubs or trees are absent from areas where one might expect them to be present.

Also, because of the prolonged summer drought in much of the state, plants that are unable to survive long periods without rainfall do not become established. In general, winter rains are more or less dependable, but there are some notable exceptions to this reliability. For example, in the winter of 1850-51 San Francisco received only slightly over 7 inches (18 cm) of rain, which is about one-fourth of its average rainfall. Clearly, that year was a "bad" one for the plants! Also, prolonged drought during the winter may also exert a negative influence on plants, even though early winter and late winter rains may produce an average total amount of rain for the year. It is these seasonal or yearly bottlenecks in rainfall that have a very important local effect on plants.

As mentioned above, the rain in California tends to be highly seasonal and the summers are generally very dry and rainless. In northern California, the occurrence of heavy fogs along the coast has a two-fold beneficial effect in alleviating some of the effects of summer drought. One of these effects is that the fogs reduce the amount of water loss from plants and from the soil so that what little water there is can be conserved. A second effect is seen in the familiar "fog drip" that occurs from the foliage and branches of tall trees along the coast. This is particularly noticeable in redwood forests, but also occurs in other coniferous forests, as well as in eucalyptus groves in some areas. The fog that condenses on the upper portions of the trees drips down to the soil and in some areas has been estimated to be equivalent to an extra ten inches of rainfall per year.

Snow is also important as a source of moisture for plants and serves as an important insulating agent for many plants in alpine areas that have severe cold and strong winter winds. Snowfall in the winter can be expected in the Sierra Nevada at any elevation above 2,000 feet (610 m). Above 4,000 feet (1,220 m) the snow may remain on the ground for long periods of time, and at higher elevations snow remains on the ground during the entire winter. The coastal region is mostly free of winter snow, although peaks in the Coast Ranges and the southern California ranges may have snow on the ground for weeks or months at a time. The snow season in the Sierra is between October and June, the actual length of time depending on the season and on the elevation.

41

Snowfall may be extremely heavy during some years. For example, at Tamarack, Alpine County, a total of 884 inches (2,245 cm) of snow was recorded during the 1906-07 winter. This is equal to almost 74 feet (22.6 m) of snow!

Although the temperature regime over much of California is moderate, extreme temperatures have been recorded for various localities. The lowest temperature recorded in the state was -45°F (-43°C) at Boca, Nevada County, which is east of Truckee. This amazingly low temperature was recorded on January 20, 1937. Since Boca is only at about 5,500 feet (1,676 m) elevation, it is quite probable that even lower temperatures have occurred in the state but have been unrecorded. The highest temperature in the state (and almost the highest temperature for any station on earth) was 134°F (57°C) in Death Valley. Both temperature extremes occurred in areas that are well vegetated, so some plant species are able to tolerate them.

The frost-free season, which agriculturalists call the growing season, varies in length from place to place. The longest growing season is 365 days along parts of the extreme southern coast of California. In the Central Valley the season is about 260 days long. In northeastern California, it is 100 to 120 days long, and at elevations of 6,000 feet (1,829 m) or above it rapidly drops off to below 100 days. There are some areas in the state, therefore, which rarely if ever experience a frost; other areas in the high montane region may have night-time temperatures that frequently drop to freezing or below even in midsummer.

The average climate of a region is important in determining what sort of vegetation and plant communities occur there. However, extreme deviations from the average climate may also have a striking effect on the plants of an area and may exert a determining role if these extremes occur frequently. For example, in parts of northern California there was a very hard freeze with unprecedented low temperatures in December, 1972. This exterminated or damaged large plantings of orchard crops and ornamentals, and also damaged a number of native trees and shrubs. Similar freezes in the southern part of the Great Basin have been known to kill or damage vast acreages of the native Creosote Bush (*Larrea divaricata*, Zygophyllaceae) at the northern edge of its range. In addition, prolonged droughts may also be effective in reducing or eliminating populations of certain perennial plants.

In general, winds in California are relatively unimportant in their influence on plant life, but in many coastal areas the persistent and occasionally very strong winds may have an effect in influencing the growth patterns of woody plants. For example, at Point Reyes, Marin County, just north of San Francisco, winds in excess of 75 miles (120 km) per hour are recorded regularly during each month from January through May. Although there are few trees on the coastward portions of Point Reyes, the trimming effect can be seen in the pine and bay forests that occupy the exposed ridges just inland from the coast. In southern California, the occasional dry, gusty "Santa Ana" winds may blow toward the coastal regions from the north or northeast. Likewise, in the Sacramento Valley, there are periods during which the dry "Northers" blow. If these strong winds occur during the growing season, they may contribute to a rapid drying of the soil which in turn results in a rather poor growth of native annuals. If these winds occur during the summer months, they considerably increase the danger of grass, brush, or forest fires and also aid in spreading fires once they become started.

At one time in the geological history of California the state had a mild, wet climate with abundant rainfall distributed throughout the year. Since Pliocene times, however, the summer season has become longer, warmer, and drier. Total rainfall has decreased and has become limited to the winter months. This increasing aridity over a long period of time resulted in striking vegetational changes in the state and was associated with the rather rapid evolution of a large number of plant species that are adapted to the modern climate of the state. At the same time that these species were evolving, a number of plant species and entire plant communities became extinct in the state.

About half of the plant communities present in California are strongly characteristic of the state and are closely adapted to its present Mediterranean climate. Some of these plant communities are restricted to California, or extend only slightly into adjacent areas. Other plant communities are adapted to wetter conditions

43

or are better developed and more extensive outside the boundaries of the state.

Adaptations to Aridity

How do California plants cope with the Mediterranean climate and its prolonged periods of summer drought, as well as with the unreliable winter rains? There are several ways in which plants have responded to this climatic regime.

Annuals are plants that complete their life cycle within a year; that is, the seeds germinate and the plants grow, flower, and set seed in less than 12 months. Many California annuals have evolved interesting mechanisms that are direct adaptations to growing in areas with a highly seasonal rainfall. Studies by a number of workers, in particular a group of biologists who worked at the California Institute of Technology in Pasadena some years ago, have investigated these adaptations. For example, the seeds of several annual species do not germinate unless they have been drenched with more than a half inch (1.3 cm) of rainfall (or its simulated equivalent in the laboratory). This water must come from above and actually wash over the seeds; placing the seeds in a bed of wet soil will not induce them to germinate. The basis of this behavior is the leaching of chemical inhibitors from the dormant seeds of these annuals, or the leaching of germination-inhibiting salts from the soil. A desert plant which germinates after the first slight rain in the autumn has a very low chance of continuing to survive and grow to maturation, and some desert annuals do not germinate immediately after the first heavy rains but exhibit a delayed germination phenomenon. The adaptive value of this trait is that such a delay, until after one or more heavy rainfalls, increases the chance that the seedlings will be growing during a period of good soil moisture. All the germination patterns that have been studied in desert annuals are explicable in terms of the average pattern of winter rains in desert areas. Obviously, any species that is unable to respond to this average pattern will have a poor chance of survival over a period of many generations.

Some of the southern California deserts receive summer rains in addition to the winter rains. It is of interest to note that in a

44

lowland cismontane non-desert station such as Berkeley, for example, the average monthly rainfall for the month of August is only 0.05 inch (0.1 cm); in Indio, in the desert, the August rainfall averages 0.38 inches (1 cm). Summer rains generally are less reliable than winter rains, and the amount of rain that falls during the summer in these desert areas usually is much lower than that which falls during the winter. As a result, some areas of the Colorado Desert (particularly that portion located in Arizona) support two somewhat different sets of annual plant species. Winter annuals germinate and grow during the winter and flower in the spring; these species provide the spectacular displays in such areas as the Anza-Borrego Desert in southern California. Less well known is the smaller number of summer annuals which germinate after summer rains and flower during summer months.

The winter annuals that have been studied germinate only when the temperatures are relatively low, thus being prevented from germinating during the summer rains. In addition, such plants will not flower until the days reach a critical length in the spring, after the cool wet winter season. These winter annuals are "informed" that spring has arrived by day-length rather than by temperature or moisture conditions, perhaps because over the long run day-length is a more reliable indicator of season than are other environmental conditions. In contrast, summer annuals germinate only at a warm temperature and thus appear during the summer but not the winter. These plants have no photoperiod requirement for flowering; they will flower when the plants have reached a suitable size for the production of flowers. Because these plants carry out their growth during the relatively benign temperature regimes that exist in the summer, they do not require a further mechanism to delay their flowering until a specific season has been reached.

Another class of plants that live in arid regions are called *phreatophytes*. Phreatophytes are perennial plants that have extensive and deep root systems that enable them to tap underground sources of water. Greasewood (*Sarcobatus vermiculatus*, Chenopodiaceae) is an example of a phreatophyte that occurs in desert regions. The young seedlings of most phreatophytes produce extensive root systems very rapidly during the winter growing season, and if these roots reach the permanent underground water supply, further growth of the plant is not directly dependent on local rainfall. However, most seedlings produced by phreatophytes are not successful in reaching ground water supplies and as a consequence seedling mortality is generally very high.

Xerophytes include succulents, which are plants such as various cacti or members of other families having fleshy stems and leaves that enable them to store water for long periods of time. Succulent xerophytes frequently have shallow root systems and thus are able to utilize the soil moisture that results from a light rainfall or from heavy dew. Such plants take advantage of what little precipitation falls in desert regions and store this water for months or years, during which time it is slowly and economically used in the metabolism of the plant. Many succulents, such as most cacti, are leafless and are so shaped that they present a minimum surface area from which water loss can occur.

Non-succulent xerophytes, such as some species of the sagebrush genus *Artemisia* (Compositae), Creosote Bush, and Ocotillo, have developed various means other than water storage in succulent tissue to endure long periods of drought. The means by which xerophytes deal with the scarce water supply vary. Some of these plants are able to obtain water from the soil even when it is present in very low amounts, because they have a high diffusion pressure deficit within their root cells, thus enabling the roots to take up what little water is present in the soil long after rains have fallen. Many of these xerophytes have developed combinations of other characteristics which enable them to economize on water. These include the presence of a heavy waxy cuticle on the leaves and stems which reduces water loss from these tissues; presence of dense mats of hairs, which have the same function; vertical orientation of leaves which places the leaves at such an angle that they receive the full sunlight obliquely rather than directly, and thus do not become heated; grayish color of leaves and stems due to pigmentation, waxes, or hairs which also reduces heating-up of plant tissues; leaves curling (or dropping) during drought periods to reduce the surface area from which water loss may occur; sunken stomates (pores) on the leaves to reduce water loss; and wide spacing of plants, perhaps in response to the low water supply. Further, a variety of thorns, spines, or essential oils may serve to discourage browsing animals from eating the leaves and stems of xerophytic plants.

Ecotypes and Life Zones

Many wide-ranging species of plants are variable throughout their range. This variability may be expressed in morphological characteristics such as height of plant or size of leaves; in "behavioral" characters such as time of flowering or season of leaf fall; or in subtle physiological characters such as tolerance to specific soil conditions, such as serpentine soil. Some environmental characters, soil type, for example, show a discontinuous distribution, and plants respond accordingly. However, one important environmental feature which shows a graded variation is that complex of phenomena which we collectively call climate. Climatic factors such as average rainfall, average temperature, etc., tend to show rather gradual changes from one area to the next. The difference in the average annual climate between a locality at 3,000 feet (914 m) in the Sierra Nevada or San Gabriel Mountains and one at 4,000 feet (1,219 m) in the same mountains is very slight. The difference between 3,000 and 5,000 feet (914 and 1,524 m) is stronger. Obviously the difference between the average climate at sea level in coastal California and the top of Mount Dana in the Sierra Nevada or Mount San Jacinto in southern California is very great. Few plant species occupy such a diversity of habitats, but there are a few plant species in California which are widely distributed and occur in a variety of climatic regimes. In view of the interesting adaptations demonstrated by soil ecotypes of such species as *Gilia capitata,* what adaptations to climatic differences can we expect in a climatically diverse plant species which occurs in a wide variety of habitats in California?

The Carnegie Group

In the 1930's, a trio of botanists in California began a long series of investigations aimed at answering the question posed above. Their results were published primarily in a monographic

47

series entitled *Experimental Studies on the Nature of Species*. The trio consisted of Jens Clausen, a Danish cytologist (a biologist who investigates chromosome number, structure, and behavior), William Hiesey, a plant physiologist, and David Keck, a plant taxonomist. These three men worked at the Carnegie Institution of Washington's Division of Plant Biology housed on the Stanford campus. The Carnegie group established three gardens which provided the growing grounds for some of their experimental plants. These gardens are located at Stanford, Mather, and Timberline.

The Stanford garden is situated in the South Coast Ranges at an elevation of about 90 feet (27 m). The natural vegetation of this area is (or was) oak savanna, which belongs to the Valley and Foothill Woodland plant community. The average growing season here is about 280 days. There is no winter snow and freezing temperatures are uncommon. The average rainfall is about 12.5 inches (32 cm) per year, most of which falls in the winter. The summer is dry and relatively warm.

The second garden, at Mather, is on the western slope of the Sierra Nevada. The elevation of this garden is somewhat over 4,000 feet (1,219 m). It is located in a well-developed coniferous forest which belongs to the Montane Forest plant community. The growing season here is 145 days long and there are moderate winter snows. The annual precipitation averages 38.5 inches (98 cm) and, although July, August, and September are generally dry, occasional rains may fall during these months.

The third garden is at a locality called Timberline, which is just east of the crest of the Sierra Nevada at an elevation of over 9,000 feet (2,743 m). This is an alpine area located in a montane meadow near the vegetational timberline. The growing season here is only 67 days long and there are heavy and prolonged snows in the winter months. Precipitation is 29 inches (74 cm) of rain and snow per year, and there is no distinct dry season during the summer.

The Carnegie group was interested in learning how certain widely distributed plant species are adapted to the climatic regime that occurs over much of California. They selected as their experimental plants some relatively widely distributed species which included the Yarrow (*Achillea*, Compositae, Plate 4A). Since the taxonomy of the *Achillea* is a bit complex, it is sufficient to say that this plant belongs to the *Achillea millefolium* group. For our purposes, we can consider that it consists of a single species in the area studied.

Seeds of *Achillea* were collected from natural populations distributed across California from the sea coast to the alpine regions. Seedlings from these wild plants were grown at Stanford and each of 60 seedlings was subdivided into three rooted cuttings. One cutting from each seedling was planted in the Stanford garden, one at Mather, and one at Timberline. The three individuals obtained in this manner from a single parent together constitute a *clone,* which is a term that refers to all individuals vegetatively propagated from a single "mother" plant. Each of the 60 seedlings was cloned, and the result was that each of the 60 individuals at Stanford had an identical mate at Mather and at Timberline. Use of the clone method enabled Clausen, Keck, and Hiesey to study the behavior of 60 "individuals" grown in three different places at once.

After a suitable time interval, the Carnegie group started the first of a series of measurements of the plants in the three gardens to determine their responses to the different climatic conditions that prevail in these three localities.

Although several characteristics were measured and several populations were studied, the following synoptical table gives measurements for three characteristics of four sample populations of *Achillea,* since this condensed version of the extensive experiments suffices for illustrative purposes. All figures given in the table are means. The term "Bodega population" refers to plants that originated near Bodega Bay, Sonoma County, an area on the immediate coast of northern California not far from San Francisco. The Clayton population came from the vicinity of Clayton, Contra Costa County, near Mount Diablo, which is at the edge of the Central Valley east of San Francisco. The Mather population originated at Mather in the vicinity of the Mather garden in the Sierra foothills. Big Horn Lake is over the crest of the Sierra Nevada at approximately 11,000 feet (3,353 m). Its climatic regime is similar to that of Timberline.

CHARACTER MEASURED	STANFORD	MATHER	TIMBERLINE
Bodega population			
Stem length ($\bar{x}$)*	48.9 cm	30.8 cm	died
Number of stems ($\bar{x}$)	19.3	18.2	---
Time of first flowers ($\bar{x}$)	May 21	July 11	---
Clayton population			
Stem length ($\bar{x}$)	70.0 cm	37.3 cm	died
Number of stems ($\bar{x}$)	16.0	7.5	---
Time of first flowers ($\bar{x}$)	April 13	June 17	---
Mather population			
Stem length ($\bar{x}$)	79.6 cm	82.4 cm	34.3 cm
Number of stems ($\bar{x}$)	7.2	28.3	0.7
Time of first flowers ($\bar{x}$)	May 15	June 30	Sept. 20
Big Horn Lake population			
Stem length ($\bar{x}$)	15.4 cm	19.5 cm	23.6 cm
Number of stems ($\bar{x}$)	2.9	3.3	3.7
Time of first flowers ($\bar{x}$)	April 29	June 6	August 14

*($\bar{x}$) = mean measurement.

Before commenting on the results given in the above table, some explanation must be given. In nature, the plants of *Achillea* native to coastal areas grow actively all year round. They occur in an area which has cool, foggy summers and cool, rainy winters. Frosts are uncommon. Because of frequent exposure to sea winds, many of the coastal plants are dwarfed in stature. In contrast, plants of *Achillea* native to the valley areas grow rapidly during the wet season of winter, and become dormant with the onset of drought in the late spring. The plants that occur naturally at Mather generally become dormant in the winter because of the cold weather, although plants from slightly lower elevations are winter-active just as are the valley plants. Plants that occupy alpine areas (such as Big Horn Lake) are dormant during the long, cold winters and during much of the winter-time are covered by heavy layers of snow. Such plants grow and flower only during the short summer months after the snow has left the ground. In nature, therefore, *Achillea* shows a diversity of responses which

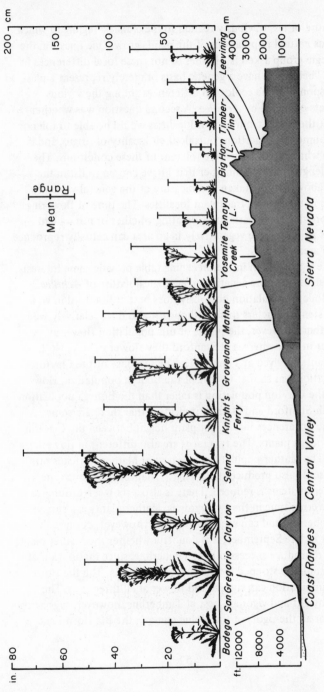

Fig. 7. Diagrammatic transect across central California showing origin of several *Achillea* populations studied by the Carnegie group. Plant specimens represent the mean height of each population grown at Stanford.

are more or less obviously related to the climatic regime in which various natural populations of the plants grow. One question the Carnegie group asked is whether or not these local differences in plant "behavior" have a genetic basis or merely represent a plastic response to the ecological differences among the various habitats occupied by *Achillea*. A second question was whether or not the plants from various localities would be able to tolerate environmental conditions outside their locality of origin, and if so, to what extent they were tolerant of these conditions. The stem length and stem number that I have chosen to list in the table above may be taken as a measure of the general vigor of plants in the various transplant localities. The time of flowering may provide some basis for estimating whether or not a plant that is able to thrive vegetatively in an area can actually reproduce in the area via seeds.

The figures given in the preceding table provide some interesting insights into the genetic and ecological nature of *Achillea*. The Bodega population obviously does best at Stanford in view of its stem characters there, although it also does relatively well at Mather. However, the plants of this population flower at Mather in July whereas at Stanford they flower in May. No Bodega plants survived at Timberline. The response of the Clayton population was similar to that of the Bodega population. However, the Clayton population is taller than the Bodega population at both Stanford and Mather, indicating that there are some genetic differences that affect plant height between the Clayton and Bodega plants. The two races are also different in flowering time. The Mather population does best at Mather, which is no surprise. It also produces tall stems at Stanford, although the number of stems is reduced. There is also a six week difference in flowering time in the two gardens. Mather plants can survive at Timberline and can even flower there; however, because they flower in late September, it is doubtful whether the Mather plants could reproduce successfully at Timberline, since maturation of seed would be stopped by the first killing frosts. The Big Horn Lake population can survive and flower at all three sites. The plant height and vigor are best at Timberline, however, suggesting that for all the rigors of the alpine climate, the Big Horn Lake plants "like" their native climate better than the relatively gentle climate of Stanford. There is a difference of four months between the time of flowering of the Big Horn Lake plants at Stanford and their time of flowering at Timberline. The early flowering at Stanford is probably a result of the fact that the plants can develop sufficient food reserves during the mild winter to produce flowers in early spring; at Timberline, the plants require about two months to produce flowers after the snow leaves the ground. Presumably flowering in August is generally early enough to develop seeds before the first killing frost of autumn, although there must be numerous years when seeds do not develop because of early severe freezes.

The observations of the Carnegie group on *Achillea* and some other, unrelated plant species that are also widely distributed in California suggest that each of these species is able to occupy a wide range of habitats not because of the great plasticity of a single genetic type, but because each of these species is subdivided into a number of local, climatically adapted, genetically different races. These climatic races are called *climatic ecotypes,* in the same way that serpentine-tolerant and -intolerant races of *Gilia capitata* are called *edaphic* (soil) *ecotypes.*

The Carnegie group concluded that in California *Achillea* is made up of eleven statistically different climatic ecotypes that occur along the 200-odd mile (322 km) transect running from the Pacific coastline to the eastern boundary of the state. They suggested that there are probably hundreds of ecotypes of *Achillea* over its entire range in Europe and North America. To paraphrase some of the major conclusions of the Carnegie workers based on their transplant work:

1. There is an intricate balance between a plant and its environment.

2. Species consist of ecotypes, each of which is in equilibrium with its environment. (However, their transplant work indicated that there was a fair amount of variability of response within each population. While the figures given in the table above are averages, they are averages of a range of responses. This variability suggested to Clausen, Keck, and Hiesey that each of the

populations of *Achillea* has the potential for responding to gradual changes to its environment, that is, these populations have not become so closely and invariably adapted to their immediate habitat that they would become extinct as a result of slight environmental change.)

3. A species is widespread only if diversified into local ecological races or ecotypes. (This generalization may not be true of weedy plant species or some forest trees.)

One practical consequence of an understanding of the genetic complexity of wide-ranging plant species is that now, when foresters choose seeds of forest trees to provide seedlings for reforestation of logged or burned areas, an attempt is made to select seeds originating from populations growing near the area to be re-seeded, or at least from a population occurring under similar ecological circumstances.

Life Zones of California

Much of the preceding discussion has been concerned with factors that affect the distribution of plant species in California. Any factor that influences the distribution of a plant species will also have an influence on the distribution of plant communities, since a plant community is an aggregation of several plant species.

An early attempt to describe the biotic composition of North America was made by the American biologist C. Hart Merriam in the 1890's. Merriam recognized that as one ascends a mountain the vegetation is stratified into horizontal zones or bands that are characteristic of certain elevations. In a very general way, these changes in the vegetation that are observed as one ascends a mountain are similar to the latitudinal vegetational changes that occur from sea level in the southern parts of North America to sea level in the northern part of the continent. For example, the tundra vegetation of mountain tops in the central Rockies bears a general similarity to the vegetation of the arctic coast of Alaska — or at least this similarity seemed to be a real one to Merriam. As a consequence of these general observations, Merriam formulated the *Life Zone* concept. The essence of this idea

54

is that North America can be divided into life zones which, according to Merriam, represent a vegetational response to temperature.

The following discussion of life zones in California is adapted from the introduction to W. L. Jepson's *Manual of the Flowering Plants of California* published in 1925. Considering the time at which Jepson's observations were published, they exhibit a remarkably modern insight into fundamental ecological principles. They also provide a valuable general, overall description of the vegetation of California.

As a result of the varying combinations of climatic factors in the state, intensified by distance from the ocean and by altitudes, the vegetation of California is markedly stratified into horizontal bands called life zones. Six life zones are recognized: 1. Lower Sonoran; 2. Upper Sonoran; 3. Transition; 4. Canadian; 5. Hudsonian; 6. Boreal. The isothermal lines of a temperature chart of California correspond in a general way, though not exactly, with these life zones, while a contour rain chart shows similar correspondence. Annual rainfall, which is slight in the deserts of the Lower Sonoran, increases one-half inch (1.3 cm) for every one hundred feet (30.5 m) in proceeding up the west slope of the Sierra Nevada. Insolation increases to the southward and also increases markedly with altitude. Humidity is greatest along the coast and diminishes toward the interior and in the south.

The Lower Sonoran Zone comprises three distinctive areas: (1) Colorado Desert or Colorado Sonoran; (b) the Mojave Desert or Mojave Sonoran; and (c) the Great Valley or Valley Sonoran. The two deserts are characterized by a typical desert climate. They have a low humidity and a low rainfall, the annual precipitation varying from 0 to about 5 inches (0 - 13 cm). They have high summer temperatures, averaging from 90° to 130°F (32° to 54°C); they have low winter temperatures, varying from about 15° to 50°F (- 9° to 10°C); and they have a great annual temperature range and a great diurnal range. Drying winds of gale force are prevalent. The vegetation of the Lower Sonoran Zone has the characteristic aspect of plants of desert regions, that is, there is everywhere exhibited a marked development of structures to inhibit transpiration or of physiological devices for the conservation of water.

The Colorado Desert lies at a low altitude, between 0 and about 500 feet (152 m). Characteristic species are members of the Goosefoot family (Chenopodiaceae), a few low bushes, and several herbs. Only a few species of trees occur, such as Desert Ironwood (*Olneya tesota*), Mesquite (*Prosopis juliflora* var. *torreyana*), and Screw Bean (*P. pubescens*), these being limited to stream beds and the borders of springs or low lying valleys. This desert passes gradually into the Mojave Desert.

The Mojave Desert lies at a higher level than the Colorado Desert, the altitudes ranging from about 2,000 to 5,000 feet (610 to 1,524 m), and the rainfall is usually somewhat greater. In other respects it has a desert climate similar to that of the Colorado Desert, and its vegetation presents a similar desert aspect. Hundreds of square miles exhibit the dark green of Creosote Bush (*Larrea divaricata*), a shrub commonly 3 to 6 feet (0.9 to 1.8 m) high with very small resin-covered evergreen leaves, the individuals widely but rather regularly spaced in response to the meagreness of soil-water (or perhaps allelopathy). Low shrubs or bushes of gray hue are abundant, and include various widely distributed species of the desert valleys belonging to the salt-bush genus *Atriplex,* and several members of the Compositae. Bladder Sage (*Salazaria mexicana*) and box thorn (*Lycium andersonii* and *L. cooperi*) are roughish or spiny shrubs. Seep weed (*Suaeda* spp.) is characteristic of alkaline valleys, while Turpentine Broom (*Thamnosma montana*) is a small switch plant of the arid slopes. There are several characteristic desert herbs as well. Extensive groves of Joshua Tree (*Yucca brevifolia*), the individuals 16 to 30 feet (5 to 9 m) high, lend an added touch of strangeness to the xerophytic populations. Save for this one species, true trees are mainly absent except that along stream courses, about springs, or in low valleys where roots may go down 20 to 70 feet (6 to 21 m) to a low-lying water stratum, a few species occur such as the mesquites *Prosopis juliflora* var. *glandulosa* and *P. pubescens*.

The Valley Sonoran comprises the plain of the Central Valley of California except the lower or central delta portion. It is a grassland formation, varying in altitude from 10 to 500 feet (3 to 152 m), with less extremes of temperature than the desert areas and a greater rainfall. In its primitive condition it is characterized by vast numbers of annuals which germinate with the winter rains and flower during the vernal period; characteristic perennial herbs are California Poppy (*Eschscholzia californica*) and Gum Plant (*Grindelia camporum*). Large areas of alkaline flats are encountered, especially on the west side of the valley. On the valley floors or undulating plains the traveler finds small depressions, a few meters square and a few centimeters deep, which fill with water in the rainy season. When a little deeper, well-defined and numerous, they take the name of "hog-wallows". With the coming of the dry season the water evaporates, and the beds of these pools in late spring or early summer give rise to a distinctive flora composed of a number of species of mostly annual plants. The narrow curtain of trees along the streams is composed of California Sycamore (*Platanus racemosa*), Fremont Poplar (*Populus fremontii*), and a few willows — though the willows are not confined to the valley floors.

The Upper Sonoran Zone may often be divided into the lower foothill belt and the chaparral belt. The lower foothill belt is a grassland formation, sometimes with a scattered growth of Blue Oak (*Quercus douglasii*) and Engelmann Oak (*Q. engelmanii*), plus several characteristic herbs. Above the lower foothill belt is the chaparral belt, or hard chaparral. It has an average altitude of 100 to 400 feet (30 to 122 m) and is characterized by the presence of extensive brush lands. Most of the species represent extreme arid-land types and possess various markedly xerophytic structures, such as small or reduced leaves, entire leaves, thickened epidermis, hard and very dense wood, vertically placed leaves, small flowers, and seeds adapted to xerophytic conditions. The most widely spread and characteristic species are three species of California lilac (*Ceanothus*), several manzanitas (*Arctostaphylos* spp.), and Mountain Mahogany (*Cercocarpus betuloides*). Many of the species of Chaparral inhabit rocky or gravelly slopes or ridges and grow on well-drained slopes.

Coastal southern California, below about 4,000 or 5,000 feet (1,219 or 1,524 m), lies mostly in the Upper Sonoran Life Zone. Some of the more important species are: Goldenbowl Mariposa (*Calochortus concolor*), and the related *C. plummerae*; Engelmann

57

Oak (*Quercus engelmannii*); the buckwheat *Oxytheca trilobata*; *Clematis pauciflora*; Matilija Poppy (*Romneya coulteri*); the jewel flower *Streptanthus campestris;* Laurel Sumac (*Rhus laurina*) and Lemonadeberry (*R. integrifolia*); *Ceanothus megacarpus* and Red-heart (*C. spinosus*); and the monkey flower *Mimulus brevipes*.

The Transition Life Zone is well defined, especially on its lower borders. The Sierra Transition lies between average altitudes of 2,000 and 5,000 feet (610 and 1,524 m) and has a mean temperature of 55° to 60°F (13° to 16°C) and an average rainfall of 25 to 35 inches (64 to 89 cm). It includes the main forest belt, and is repeated in the mountains of Southern California and in the Coast Ranges, where the latter rise to sufficiently high altitudes. This life zone is distinctive and, on the whole, rather definitely circumscribed. It contains, for California, a greater number of species of trees and shrubs than any other life zone and has, in addition, a very large population of herbs.

Widely developed in some parts of the state and very narrow in others, the Arid Transition of the Great Basin underlies the lower margin of the main Humid and Sierra Transition. It is, in California, a drier and more exposed sub-area, often with a preponderance of brush slopes and scattered trees. Its most characteristic species are Ponderosa Pine (*Pinus ponderosa*), Basin Sagebrush (*Artemisia tridentata*), the manzanita *Arctostaphylos patula, Garrya fremontii* and the plum *Prunus subcordata*.

The Sierra Transition is developed and forms a broad band. It carries the less open part of the forest belt. The dominant forest species are Ponderosa Pine (*Pinus ponderosa*) and Sugar Pine (*P. lambertiana*); Incense Cedar (*Calocedrus decurrens*) and White Fir (*Abies concolor*). The Big Tree (*Sequoiadendron giganteum*) is a marked feature of this zone in the southern part of the Sierra Nevada. The dry or more open forest or forestless slopes present many shrubs of wide range such as California Hazelnut (*Corylus cornuta* var. *californica*), Thimble Berry (*Rubus parviflorus*), Service Berry (*Amelanchier alnifolia*), three species of *Ceanothus*, Mountain Dogwood (*Cornus nuttallii*), and a few other shrubby species. On the dry flats or open wet meadows or swamps as well as on the forest slopes there are various characteristic herbs.

58

The Redwood Transition, which comprises the coastal Redwood belt, extends from sea-level to 2,000 or sometimes to 3,000 feet (610 to 914 m) in altitude. It has, therefore, a very much lower altitude than the Sierra Transition and a somewhat higher rainfall, varying from 25 to 122 inches (64 to 310 cm). From the standpoint of temperature it enjoys a lower annual range and lower diurnal range. Being wholly within the coastal fog belt and lying next to the ocean, it has much greater humidity. Proceeding from the Central Valley to the coastal edge of the Redwood belt, one passes through formations similar, in an ecological view, to those met with in ascending the Sierra from the Central Valley, since there are met successively the dry barren plains, the barren foothills, the Chaparral, and finally a narrow band of *Pinus ponderosa* and *Quercus kelloggii*, which is characteristic of the Arid Transition. In its greatest development, Coast Redwood forms pure stands. In other parts of the belt it is dominant but with it are associated Tan Oak (*Lithocarpus densiflora*) and Douglas Fir (*Pseudotsuga menziesii*) and other conifers. On the inner side of the Redwood belt is a marked band of Madrone (*Arbutus menziesii*) and Douglas Fir. Herbs include Fetid Adder's Tongue (*Scoliopus bigelovii*); *Clintonia andrewsiana*; False Lily-of-the-Valley (*Maianthemum dilatatum*); *Trillium ovatum*; Ginger Root (*Asarum caudatum*); Inside-out Flower (*Vancouveria parviflora*); Vanilla Leaf (*Achlys triphylla*); *Saxifraga mertensiana*; and Redwood Sorrel (*Oxalis oregana*). Certain shrubs form a very low understory and often occur in heavy stands, as Blue Brush (*Ceanothus thyrsiflorus*); Evergreen Huckleberry (*Vaccinium ovatum*); and Salal (*Gaultheria shallon*).

The Canadian Life Zone is not well defined in the Sierra Nevada and has, as a separate zone, only a shadowy or wavering existence. Its natural place is on the average between 5,000 and 7,000 feet (1,524 and 2,134 m) where there is a mean annual temperature of 50° to 55°F (10° to 13°C) and an average rainfall of 40 to 50 inches (102 to 127 cm). The most useful indicator species in this life zone are Red Fir (*Abies magnifica*), Jeffrey Pine (*Pinus jeffreyi*), Western White Pine (*P. monticola*), and Lodgepole Pine (*P. murrayana*). The first three species are, however, often found in the upper part of the Transition Zone, associated with typical Transition species, while the fourth is frequently a characteristic species of the next life zone above, the Hudsonian. The following herbs and shrubs may be considered as belonging to this zone: Bitter Cherry (*Prunus emarginata*); Pinemat Manzanita (*Arctostaphylos nevadensis*); *Nama lobbii*; *Hesperochiron californicus*; the monkey flower *Mimulus lewisii*; and the lousewort *Pedicularis semibarbata*.

The Hudsonian Life Zone is the timber-line zone, and is fairly well defined as to its upper borders. It has an average altitude of 7,000 to 9,000 feet (2,134 to 2,743 m), a mean annual temperature of 45° to 50°F (7° to 10°C), and an average rainfall of 50 to 55 inches (127 to 140 cm). The most important index species are Whitebark Pine (*Pinus albicaulis*), Western Juniper (*Juniperus occidentalis*), and various herbaceous species.

The Boreal Life Zone is true alpine. Its altitudinal range varies from 9,000 to 14,500 feet (2,743 to 4,420 m) with a mean annual temperature of 40° to 45°F (4° to 7°C) and an average rainfall of 60 to 70 inches (152 to 178 cm). The characteristic species are almost exclusively herbs. The Boreal Zone presents marked plant formations on the Salmon Mountains, Mt. Shasta, Lassen Peak, and the high Sierra Nevada. It recurs (though represented by few species) on Mt. San Gorgonio in the San Bernardino Mountains and feebly on Mt. San Jacinto in the San Jacinto Mountains.

Merriam's classification of life zones in North America was an early attempt to subdivide the biota of the continent in a biologically meaningful manner. However, his assumption of the all-controlling influence of temperature conditions is a naive one in view of the other variables that affect plant distribution. In other words, Merriam's scheme was too simple in its approach, although it was a valuable pioneer attempt to classify the vegetation of North America according to ecological principles.

A. Desert Candle (*Caulanthus inflatus*), southern California.

B. *Fremontodendron californicum* subsp. *decumbens*, Eldorado County.

C. *Iris munzii*, Tulare County.

D. Ione Manzanita (*Arctostaphylos myrtifolia*), Amador and Calaveras counties.

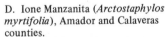

Plate 1. CALIFORNIA ENDEMICS

A. *Canbya candida*, Mojave Desert.

B. *Calochortus tiburonensis*, Marin County.

C. Digger Pine (*Pinus sabiniana*), Foothill Woodland.

D. California Buckeye (*Aesculus californica*), mostly Foothill Woodland.

Plate 2. CALIFORNIA ENDEMICS

A. *Gilia capitata.*

B. Serpentine rock with *Mimulus guttatus.*

C. Serpentine (r) and non-serpentine (l) vegetation, Sierra foothills.

D. Alkali scald with *Lasthenia chrysantha.*

Plate 3. ADAPTATIONS

A. Yarrow (*Achillea millefolium*).

C. Guano endemism: *Lasthenia minor* subsp. *maritima.*

B. Crown sprouting of Chamise (*Adenostoma fasciculatum*) after fire.

D. Fire annual: *Phacelia brachyloba.*

Plate 4. ADAPTATIONS

A. Coastal Strand.

B. Coastal Prairie.

C. Coastal Salt Marsh.

D. Dodder (*Cuscuta salina*) parasitic on Pickleweed (*Salicornia* sp.) in Coastal Salt Marsh.

Plate 5. PLANT COMMUNITIES

A. Northern Coastal Scrub.

B. Closed-Cone Pine Forest with Bishop Pine (*Pinus muricata*).

C. Knobcone Pine (*Pinus attenuata*) in Closed-Cone Pine Forest.

D. Pygmy Forest in Closed-Cone Pine Forest.

Plate 6. PLANT COMMUNITIES

A. North Coastal Forest, Coast Redwood and Douglas Fir (*Pseudotsuga menziesii*).

B. Stump sprouting, Coast Redwood (*Sequoia sempervirens*).

C. Madrone (*Arbutus menziesii*) in North Coastal Forest.

D. Salal (*Gaultheria shallon*) in North Coastal Forest.

Plate 7. PLANT COMMUNITIES

A. California Bay (*Umbellularia californica*) in North Coastal Forest.

B. Vine Maple (*Acer circinatum*) in North Coastal Forest.

C. Chaparral, North Coast Ranges.

D. Chaparral, Sierra foothills.

Plate 8. PLANT COMMUNITIES

A. Valley and Foothill Woodland (spring).

B. Valley and Foothill Woodland (late summer).

C. *Lupinus nanus* in Valley and Foothill Woodland.

D. Valley Grassland with vestiges of bunch grasses.

Plate 9. PLANT COMMUNITIES

A. Spring annuals in Valley Grassland.

B. Vernal pool in Valley Grassland.

C. Riparian Woodland.

D. Riparian Woodland.

Plate 10. PLANT COMMUNITIES

A. Freshwater Marsh.

B. Montane Forest with Jeffrey Pine (*Pinus jeffreyi*).

C. Montane Forest with Red Fir (*Abies magnifica*).

D. Sierra Big Tree (*Sequoiadendron giganteum*) in Montane Forest.

Plate 11. PLANT COMMUNITIES

A. Montane Chaparral

B. Succession: Fir (*Abies*) seedlings in Montane Chaparral.

C. Subalpine Forest with Mountain Hemlock (*Tsuga mertensiana*).

D. Timberline, Subalpine Forest.

Plate 12. PLANT COMMUNITIES

A. Montane Meadow with Corn Lily (*Veratrum* sp.).

B. Alpine Fell-field.

C. *Phlox diffusa* in Alpine Fell-field.

D. *Eriogonum kennedyi* in Alpine Fell-field.

Plate 13. PLANT COMMUNITIES

A. Pinyon-Juniper Woodland.

B. Sagebrush Scrub.

C. Coastal Sage Scrub.

D. Shadscale Scrub.

Plate 14. PLANT COMMUNITIES

A. Dry lake, Alkali Sink Scrub.

B. Alkali Sink Scrub.

C. Iodine Bush (*Allenrolfea occidentalis*) in Alkali Sink Scrub.

D. Joshua Tree Woodland.

Plate 15. PLANT COMMUNITIES

A. Creosote Bush Scrub.

B. Jumping Cholla (*Opuntia* sp.) in Creosote Bush Scrub.

C. *Oenothera deltoides* in Creosote Bush Scrub.

D. Desert wash in Creosote Bush Scrub.

Plate 16. PLANT COMMUNITIES

4. AN INTRODUCTION TO
CALIFORNIA PLANT COMMUNITIES

California Plant Communities and Their Major Components

A plant *community* is a regional assemblage of interacting plant species characterized by the presence of one or more dominant species. The concept of the community has been the subject of considerable argument in past decades and there is no uniform application of the term even today. Some botanists define a plant community simply as an assemblage of plants living in a prescribed area or physical habitat. Other botanists deny the "reality" of plant communities and do not believe that they exist, except in the minds of some ecologists. Nevertheless, there are practical reasons for recognizing plant communities in California as a basis for discussing the plant life of the state.

In Munz' *A California Flora* eleven vegetation types and twenty-nine plant communities are recognized for California, based on a scheme that Munz and D. D. Keck devised ten years earlier. The vegetation types they recognize in California are:

1. Strand
2. Salt Marsh
3. Freshwater Marsh
4. Scrub
5. Coniferous Forest
6. Mixed Evergreen Forest
7. Woodland-Savanna
8. Chaparral
9. Grassland
10. Alpine Fell-Field
11. Desert Woodland

Another classification of California plant communities that is relatively simple and useful is given below. Some of the characteristic plant species of each community are listed along with their distribution in that plant community in California. Each of these communities is discussed later in the text. This classification of California plant communities is my modification of one brought to my attention by J. R. Haller of the University of California, Santa Barbara. The phrase in parentheses under the names of the communities in the listing indicates their equivalent according to the Munz and Keck treatment. The distribution of these plant communities in California is given in Maps 2-4.

Coastal Strand: (same in Munz)

Abronia spp.*	Sand Verbena	Nyctaginaceae	widespread
Ambrosia chamissonis	Silver Beachweed	Compositae	widespread
Atriplex spp.	Saltbush	Chenopodiaceae	widespread
Camissonia (Oenothera) cheiranthifolia	Beach Primrose	Onagraceae	widespread
Lupinus arboreus and others	Bush Lupine	Leguminosae	widespread

Coastal Prairie: (same in Munz)

Various genera of grasses		Gramineae	
Chrysopsis villosa var. *bolanderi*	Golden Aster	Compositae	N. Cal. only
Iris douglasiana	Douglas Iris	Iridaceae	widespread
Pteridium aquilinum	Bracken Fern	Pteridaceae	widespread
Sanicula arctopoides	Yellow Mats	Umbelliferae	N. Cal. only

Coastal Salt Marsh: (same in Munz)

Distichlis spicata	Salt Grass	Gramineae	widespread
Frankenia grandifolia	Frankenia	Frankeniaceae	widespread
Salicornia spp.	Glasswort, Pickle-weed	Chenopodiaceae	widespread
Suaeda californica	Seep Weed	Chenopodiaceae	widespread

Northern Coastal Scrub: (same in Munz)

Anaphalis margaritacea	Pearly Everlasting	Compositae	widespread
Artemisia suksdorfii	Suksdorf's Sage-brush	Compositae	N. Cal. only
Baccharis pilularis var. *consanguinea*	Coyote Brush, Chaparral Broom	Compositae	widespread
Erigeron glaucus	Seaside Daisy	Compositae	widespread
Eriogonum latifolium	Coastal Eriogonum	Polygonaceae	widespread
Eriophyllum staechadifolium	Seaside Woolly Sunflower	Compositae	N. Cal. only
Gaultheria shallon	Salal	Ericaceae	widespread
Heracleum lanatum	Cow Parsnip	Umbelliferae	mostly N. Cal.
Rubus vitifolius	California Black-berry	Rosaceae	mostly N. Cal.

*spp.=species in the plural, abbreviated.

Closed-Cone Pine Forest: (same in Munz)

Cupressus spp.	Cypress	Cupressaceae	scattered
Pinus contorta	Beach Pine	Pinaceae	N. Cal. coast
Pinus muricata	Bishop Pine	Pinaceae	widely scattered on coast
Pinus radiata	Monterey Pine	Pinaceae	scattered, central coast

Associated Species:

Arctostaphylos spp.	Manzanita	Ericaceae	widespread
Baccharis pilularis var. *consanguinea*	Coyote Brush, Chaparral Broom	Compositae	mostly N.Cal.
Myrica californica	Wax Myrtle	Myricaceae	mostly N.Cal.
Pteridium aquilinum	Bracken Fern	Pteridaceae	widespread
Quercus agrifolia	Coast Live Oak	Fagaceae	widespread
Rhamnus californica	Coffeeberry	Rhamnaceae	widespread
Rhus diversiloba	Poison Oak	Anacardiaceae	widespread
Vaccinium ovatum	California Huckleberry	Ericaceae	mostly N.Cal.

North Coastal Forest: (includes: North Coastal Coniferous Forest; Redwood Forest; Douglas Fir Forest; and Mixed Evergreen Forest)

Abies grandis	Lowland or Grand Fir	Pinaceae	N. Cal. only
Acer macrophyllum	Bigleaf Maple	Aceraceae	widespread
Chamaecyparis lawsoniana	Lawson Cypress, Port Orford Cedar	Cupressaceae	N. Cal. only
Chrysolepis (Castanopsis) chrysophylla	Giant Chinquapin	Fagaceae	N. Cal. only
Lithocarpus densiflora	Tanbark Oak	Fagaceae	mostly N.Cal.
Picea sitchensis	Sitka Spruce	Pinaceae	N. Cal. only
Pseudotsuga menziesii	Douglas Fir	Pinaceae	N. Cal. only
Sequoia sempervirens	Redwood	Taxodiaceae	N. Cal. only
Thuja plicata	Canoe or W. Red Cedar	Cupressaceae	N. Cal. only
Tsuga heterophylla	Western Hemlock	Pinaceae	N. Cal. only

Understory species:

Acer circinatum	Vine Maple	Aceraceae	N. Cal. only
Gaultheria shallon	Salal	Ericaceae	N. Cal. only
Oxalis oregana	Redwood Sorrel	Oxalidaceae	N. Cal. only
Polystichum munitum	Sword Fern	Aspidiaceae	N. Cal. only
Rhododendron macrophyllum	California Rose Bay	Ericaceae	N. Cal. only

63

Rubus parviflorus	Thimble Berry	Rosaceae	mostly N.Cal.
Vaccinium ovatum and others	Huckleberry	Ericaceae	mostly N.Cal.

Drier margins of the North Coastal Forest (Mixed Evergreen Forest):

Acer macrophyllum	Bigleaf Maple	Aceraceae	widespread
Arbutus menziesii	Madrone	Ericaceae	mostly N.Cal.
Lithocarpus densiflora	Tanbark Oak	Fagaceae	mostly N.Cal.
Quercus chrysolepis	Canyon Oak	Fagaceae	mostly N.Cal.
Quercus garryana	Oregon or Garry Oak	Fagaceae	N. Cal. only
Quercus kelloggii	Black Oak	Fagaceae	widespread
Quercus wislizenii	Interior Live Oak	Fagaceae	widespread
Rhus diversiloba	Poison Oak	Anacardiaceae	widespread
Umbellularia californica	California Bay	Lauraceae	widespread

Chaparral (Hard Chaparral): (same in Munz)

Adenostoma fasciculatum	Chamise, Greasewood	Rosaceae	widespread
Arctostaphylos spp.	Manzanita	Ericaceae	widespread
Ceanothus spp.	California Lilac	Rhamnaceae	widespread
Cercocarpus betuloides	Mountain Mahogany	Rosaceae	widespread
Heteromeles arbutifolia	California Holly, Toyon	Rosaceae	widespread
Prunus ilicifolia	Holly-leaf Cherry	Rosaceae	widespread
Quercus dumosa	Scrub Oak	Fagaceae	widespread
Rhus diversiloba	Poison Oak	Anacardiaceae	widespread
Rhus laurina	Laurel Sumac	Anacardiaceae	S. Cal. only
Rhus ovata	Sugar Bush	Anacardiaceae	S. Cal. only
Yucca whipplei	Yucca, Spanish Bayonet	Agavaceae	S. Cal. only

Valley and Foothill Woodland: (includes: Northern, Southern Oak Woodland; and Foothill Woodland)

Aesculus californica	California Buckeye	Hippocastanaceae	widespread
Juglans californica	S. Cal. Walnut	Juglandaceae	S. Cal. only
Pinus sabiniana	Digger Pine	Pinaceae	widespread
Quercus agrifolia	Coast Live Oak	Fagaceae	widespread
Quercus douglasii	Blue Oak	Fagaceae	N. Cal. only
Quercus engelmannii	Engelmann Oak	Fagaceae	S. Cal. only
Quercus garryana	Oregon or Garry Oak	Fagaceae	N. Cal. only

Quercus lobata	Valley Oak	Fagaceae	mostly N.Cal.
Quercus wislizenii	Interior Live Oak	Fagaceae	widespread

Understory consists of Valley Grassland species and occasional Chaparral species

Valley Grassland: (same in Munz)

Native:

Aristida (many species)	Three-Awn	Gramineae	widespread
Poa (many species)	Bunch Grass	Gramineae	widespread
Stipa (many species)	Needle Grass	Gramineae	widespread

Introduced:

Avena (many species)	Wild Oats	Gramineae	widespread
Bromus (many species)	Brome Grass	Gramineae	widespread
Festuca (many species)	Fescue	Gramineae	widespread

Riparian Woodland: (not in Munz)

Acer macrophyllum	Bigleaf Maple	Aceraceae	widespread
Acer negundo subsp. *californicum*	California Boxelder	Aceraceae	widespread
Alnus rhombifolia	White Alder	Betulaceae	widespread
Platanus racemosa	California Sycamore	Platanaceae	widespread
Populus fremontii	Fremont Cottonwood	Salicaceae	widespread
Populus trichocarpa	Black Cottonwood	Salicaceae	widespread
Salix (many species)	Willow	Salicaceae	widespread
Umbellularia californica	California Bay	Lauraceae	widespread

Freshwater Marsh: (same as Munz)

Carex spp.	Sedge	Cyperaceae	widespread
Scirpus spp.	Bulrush or Tule	Cyperaceae	widespread
Typha spp.	Cattail	Typhaceae	widespread

HIGH MOUNTAIN (MONTANE) REGION

Montane Forest: (mostly Yellow Pine Forest in Munz)

Abies concolor	White Fir	Pinaceae	widespread
Calocedrus decurrens	Incense Cedar	Cupressaceae	widespread
Pinus coulteri	Coulter Pine	Pinaceae	S. Cal. and S. Coast Ranges
Pinus lambertiana	Sugar Pine	Pinaceae	widespread
Pinus ponderosa	Ponderosa Pine	Pinaceae	widespread

Pseudotsuga macro-carpa	Bigcone Spruce	Pinaceae	S. Cal.
Pseudotsuga menziesii	Douglas Fir	Pinaceae	Central Cal.
Quercus chrysolepis	Canyon Oak	Fagaceae	widespread
Quercus kelloggii	California Black Oak	Fagaceae	widespread
Sequoiadendron giganteum	Giant Sequoia, Sierra Big Tree	Taxodiaceae	Sierra Nevada

Understory species:

Chamaebatia foliolosa	Mountain Misery	Rosaceae	Sierra Nevada
Pteridium aquilinum	Bracken Fern	Pteridaceae	widespread
Ribes spp.	Currant, Gooseberry	Grossulariaceae	widespread

All species from Montane Chaparral

Upper margin of the Montane Forest:

Abies magnifica	Red Fir	Pinaceae	Sierra Nevada
Pinus jeffreyi	Jeffrey Pine	Pinaceae	widespread
Pinus murrayana	Lodgepole Pine	Pinaceae	widespread

Montane Chaparral: (not in Munz)

Arctostaphylos spp.	Manzanita	Ericaceae	widespread
Ceanothus cordulatus	Snow Bush	Rhamnaceae	widespread
Chrysolepis (Castanopsis) sempervirens	Chinquapin	Fagaceae	widespread
Quercus chrysolepis	Canyon Oak	Fagaceae	mostly S. Cal.
Quercus vacciniifolia	Huckleberry Oak	Fagaceae	Central Sierra Nevada n.

Subalpine Forest: (includes: Red Fir Forest; Lodgepole Forest; Subalpine Forest; Bristlecone Pine Forest)

Abies magnifica	Red Fir	Pinaceae	Sierra Nevada
Juniperus occidentalis	Sierra Juniper	Cupressaceae	mostly Sierra Nevada n.
Pinus albicaulis	Whitebark Pine	Pinaceae	Sierra Nevada north
Pinus aristata	Bristlecone Pine	Pinaceae	high desert ranges
Pinus balfouriana	Foxtail Pine	Pinaceae	S. Sierra Nevada and Scott Mountains
Pinus flexilis	Limber Pine	Pinaceae	S. Cal. and E. Sierra Nevada
Pinus monticola	Western White Pine	Pinaceae	Sierra Nevada north
Pinus murrayana	Lodgepole Pine	Pinaceae	widespread

Populus tremuloides	Quaking Aspen	Salicaceae	Sierra Nevada
Tsuga mertensiana	Mountain Hemlock	Pinaceae	Central Sierra Nevada n.

Understory:

Artemisia spp.	Sagebrush	Compositae	widespread
Ledum glandulosum	Labrador Tea	Ericaceae	Sierra Nevada north
Penstemon spp.	Penstemon	Scrophulari- aceae	widespread
Phyllodoce breweri	Red Heather	Ericaceae	mostly Sierra Nevada n.
Ribes spp.	Currant	Grossulariaceae	widespread
Salix spp.	Willow	Salicaceae	widespread
Vaccinium spp.	Huckleberry	Ericaceae	Sierra Nevada north

Many species from Montane Meadow

Montane Meadow: (not in Munz)

Veratrum califor- nicum	Corn Lily	Liliaceae	widespread

Many species of mostly perennial grasses and sedges (especially *Carex* spp.); many species of annual and perennial broad-leaved herbs

Alpine Fell-field: (same in Munz)
Many species of perennial herbs and dwarf woody plants

Pinyon-Juniper Woodland: (includes: Northern Juniper Woodland; Pinyon-Juniper Woodland)

Cercocarpus ledi- folius	Mountain Mahogany	Rosaceae	widespread
Juniperus califor- nica	California Juniper	Cupressaceae	S. Cal. only
Juniperus occiden- talis	Sierra Juniper	Cupressaceae	widespread
Juniperus osteo- sperma	Utah Juniper	Cupressaceae	Nevada bor- der region
Pinus monophylla	Single-leaf Pinyon	Pinaceae	widespread
Purshia spp.	Antelope Brush	Rosaceae	widespread
Quercus turbinella	Desert Scrub Oak	Fagaceae	S. Mojave- Colo. Desert

All species of Sagebrush Scrub

Sagebrush Scrub: (same in Munz)

Artemisia tridentata	Basin Sagebrush	Compositae	widespread
Atriplex spp.	Saltbush	Chenopodiaceae	widespread
Chrysothamnus nauseosus	Rabbit Brush	Compositae	widespread

Coleogyne ramosis-sima	Blackbush	Rosaceae	widespread
Purshia spp.	Antelope Brush	Rosaceae	widespread
Tetradymia spp.	Cotton Thorn	Compositae	Mojave Desert north

Coastal Sage Scrub (Soft Chaparral): (same in Munz)

Artemisia californica	Coastal Sagebrush	Compositae	widespread
Baccharis pilularis var. *consanguinea*	Coyote Brush, Chaparral Broom	Compositae	widespread
Eriogonum fascicu-latum	Wild Buckwheat	Polygonaceae	widespread
Rhus diversiloba	Poison Oak	Anacardiaceae	widespread
Rhus integrifolia	Lemonadeberry	Anacardiaceae	S. Cal. only
Salvia leuco-phylla	Purple or White-leaved Sage	Labiatae	mostly S. Cal.
Salvia mellifera	Black Sage	Labiatae	widespread

Shadscale Scrub: (same in Munz)

Artemisia spinescens	Spiny Sagebrush	Compositae	Mojave Desert north
Atriplex spp.	Saltbush, Shadscale	Chenopodiaceae	widespread
Coleogyne ramosis-sima	Blackbush	Rosaceae	widespread
Ephedra spp.	Mormon Tea	Ephedraceae	widespread
Eurotia lanata	Winter Fat	Chenopodiaceae	Mojave Desert north
Grayia spinosa	Hop Sage	Chenopodiaceae	Mojave Desert north
Gutierrezia spp.	Matchweed	Compositae	Mojave Desert north
Hymenoclea salsola	Cheese Bush	Compositae	widespread

Alkali Sink Scrub: (same in Munz)

Allenrolfea occiden-talis	Iodine Bush	Chenopodiaceae	widespread
Atriplex spp.	Saltbush	Chenopodiaceae	widespread
Salicornia spp.	Pickleweed	Chenopodiaceae	widespread
Sarcobatus vermicu-latus	Greasewood	Chenopodiaceae	widespread
Suadea spp.	Seep Weed	Chenopodiaceae	widespread

Joshua Tree Woodland: (same in Munz)

Atriplex spp.	Saltbush	Chenopodiaceae	widespread
Ephedra spp.	Mormon Tea	Ephedraceae	widespread
Eriogonum fascicu-latum	Wild Buckwheat	Polygonaceae	widespread

68

Haplopappus spp.	Bristlewood	Compositae	widespread
Juniperus californica	California Juniper	Cupressaceae	widespread
Lycium spp.	Box Thorn	Solanaceae	widespread
Opuntia spp.	Cholla, Prickly Pear	Cactaceae	widespread
Salazaria mexicana	Bladder Sage	Labiatae	widespread
Tetradymia axillaris	Cotton Thorn	Compositae	Mojave Desert north
Yucca brevifolia	Joshua Tree	Agavaceae	Mojave Desert
Yucca schidigera	Mojave Yucca	Agavaceae	Mojave-Colo. Desert

Most species of Shadscale Scrub

Creosote Bush Scrub: (same in Munz)

Encelia farinosa	Brittle Bush	Compositae	mostly Colo. Desert
Fouquieria splendens	Ocotillo	Fouquieriaceae	Colo. Desert
Franseria dumosa	Burro Weed	Compositae	very widespread
Hymenoclea salsola	Cheese Bush	Compositae	widespread in washes
Larrea divaricata	Creosote Bush	Zygophyllaceae	very widespread
Opuntia spp.	Cholla, Prickly Pear	Cactaceae	widespread on rocky slopes

Ecological Dominance

Some plant communities are named for the tree or shrub species which are *dominant* in them. The term dominant refers to one or more plant species which may be the largest or most abundant plants in a community, or those which account for the greatest coverage in the community. Because of the foliage cover or the extent of their root systems, dominants have a strong influence on the local ecology of the community of which they are members. Perhaps the most straightforward and familiar example of the idea of dominance is that which exists in the Redwood Forest, which is recognized by Munz and Keck as a distinct community although I have included it in the North Coastal Forest plant community. This plant association is named after its sole dominant, Coast Redwood (*Sequoia sempervirens*). Because of the large size of these trees and the influence that they have on the moisture and shading relationships under them, redwoods exert a strong influence on determining which plants can grow under them. Redwood forests tend to be densely shaded at the ground level and also to have thick mats of semi-decomposed fallen redwood leaves and twigs. Because of the shading and the dense cover of fallen leaves, rather few herbaceous plants can flourish under redwood trees, Plate 7. Among these are Redwood Sorrel (*Oxalis oregana*), Inside-out Flower (*Vancouveria parviflora*), Sword Fern (*Polystichum munitum*) and a few shrubs such as Salal (*Gaultheria shallon*), Evergreen or California Huckleberry (*Vaccinium ovatum*), and some other species.

Certain features of the Redwood Forest also support the contention of some plant ecologists that in many instances the idea of "community" is rather arbitrary. Obviously, wherever you encounter a large natural stand of redwoods you are, by definition, in a redwood forest. A redwood forest is defined by the presence of a single dominant species. One might assume, however, that a number of "fellow travellers" of the Coast Redwood also are generally restricted to redwood forest. But this is not true. If you examine the distribution of all other plant species that generally are associated with redwoods you will find that each of these is widespread in areas where the Coast Redwood does not occur, and, furthermore, that there are no striking parallelisms among the ranges of these individual species.

Examples of other Coniferous Forest plant communities dominated by one or a few tree species are the Closed-Cone Pine Forest (dominated in various areas by Bishop Pine, *Pinus muricata*; or Monterey Pine, *P. radiata*) and the Montane Forest (dominated in various areas by the Ponderosa Pine, *Pinus ponderosa*; Coulter Pine, *P. coulteri*; Sugar Pine, *P. lambertiana*; Incense Cedar, *Calocedrus decurrens*; Douglas Fir, *Pseudotsuga menziesii*; and/or a few other coniferous tree species).

The aggregations of plants in some parts of California cannot be recognized as belonging to any plant community. In many areas, there are mixtures of species that are characteristic of two or more different plant communities. One solution to this problem is to continue to invent new names to designate these associations, but ultimately this would lead to a confusingly large number of plant communities. Perhaps the most reasonable alternative is to remember that plant communities are defined by man, not by nature, and plants seldom follow the rather rigid limitations set down by man. The variety in the plant cover of California is a reflection of the richness of the flora of the state and of the remarkable diversity of ecological conditions that interact to influence the patterns of distribution of each plant species.

Succession

Most of the plant communities that occur in California represent *climax* plant communities. The idea of climax goes back to Frederick Clements, an American ecologist who invented the term because of its similarity to the word climate. As a midwesterner it was probably natural that Clements believed that climate was the chief determining influence in the stable vegetation that occurs in any area, but in California other influences (such as soil type) may be extremely important in influencing the distribution of plants. A climax plant community is the final, self-perpetuating plant community that will occur in an area under stable ecological conditions. Examples of climax plant communities in California are Valley and Foothill Woodland, Creosote Bush Scrub, and Valley Grassland.

However, whenever a plant community is severely disturbed by fire, glaciation, or other ecological influences, it may not be re-placed immediately by re-establishment of the same plant species that were present before the disturbance. For example, the first plants to occupy an area in California after a hot forest fire are generally not seedlings of the tree, shrub, or herbaceous species that previously occupied the site, but are seedlings of other species that typically occur after fires sweep through an area. In the upper-middle Montane Forest of the western slope of the Sierra Nevada, there is a characteristic forest flora which is subject to frequent burning. After the fires sweep through these forests, the first plants to appear in the following year are various herbaceous annuals and perennials. Later, shrubs become established and eventually provide the shade that is necessary for the establishment of seedlings of the climax forest trees (Plate 12B). Seedlings of pines and firs become established under these shrubs

71

and eventually grow up through the shrub layer. After a period of several years the large trees cast sufficient shade to inhibit further establishment of shrub seedlings and to allow the establishment of seedlings of additional forest trees.

Ultimately the plant community reaches a stage where it does not undergo a directional change in species composition, but has achieved a stability that is reflected by the fact that the youngest plants in the area are the same species as the oldest ones. The change in the plant communities that may occur on a given site over a period of time is called *succession*. Succession may occur after gross ecological disturbances, or may result from more subtle changes such as the gradual filling in of a pond with the consequent replacement of an aquatic community by a terrestrial one, or from the occupation of a bare rock surface by lichens, mosses, and ultimately other plants.

In areas which are subjected to continual burning, a climax plant community may never develop, but the site may be occupied by a *subclimax* community. For example, Douglas Fir is not a climax tree in many areas west of the Cascades in the Pacific Northwest, but forests of this species are maintained over long periods of time in this region because logging activities make conditions suitable for establishment of Douglas Fir seedlings but not of seedlings of the coniferous species that are components of the climax vegetation. Likewise, repeated fires in many areas of the western United States prevent the development of climax plant communities. Plant communities that are not climax, therefore, cannot be preserved by maintaining a policy of "hands off." For example, in many (but not all) areas of California, Chaparral is maintained by periodic fires. Conservationists must appreciate the fact that some western plant species and plant communities are more likely to disappear if they are undisturbed than if they are allowed to burn occasionally.

Another kind of climax that may develop is the *disclimax*. In areas of the Central Valley that were occupied by native perennial grasslands, the native species have disappeared and have been replaced by introduced annual grasses which are maintained because of grazing. In this case, an unnatural plant community has been induced which is not a component of the natural succession-

72

al changes characteristic of the area. If grazing were removed from these areas, they would perhaps eventually revert to perennial grassland, although in many areas the native grasses are completely extinct and therefore cannot function as a source of seed. Under these circumstances, it is uncertain what successional changes would take place if grazing animals were removed.

5. A TRANSECT ACROSS NORTHERN CALIFORNIA

The following pages present a synoptic account of the composition and distribution of California plant communities, together with some discussion of outstanding characteristics of the communities or of certain of their component species. In northern California, all major plant communities of the North Coast ranges and adjacent coastal areas can be encountered by making a transect starting from the shores of the Pacific Ocean in Marin County and working eastward across the North Coast ranges, the Central Valley, and the Sierra Nevada to the Nevada state line.

Coastal Strand (Plate 5A, Map 2)

The first terrestrial plant community encountered above the high tide line in northern California is the Coastal Strand community. Common plants in the Coastal Strand community are *Ambrosia chamissonis,* the Silver Beachweed of the Compositae; various species of the saltbush genus *Atriplex* (Chenopodiaceae); various lupines such as the beautiful yellow Bush Lupine (*Lupinus arboreus,* Leguminosae); the colorful sand verbena (*Abronia* spp., Nyctaginaceae); Beach Primrose, *Camissonia cheiranthifolia,* of the Onagraceae. Also present are the introduced succulent ice plants *Mesembryanthemum nodiflorum* and *M. crystallinum* (both Aizoaceae); and the attractive Beach Morning Glory, *Convolvulus* (sometimes called *Calystegia*) *soldanella* (Convolvulaceae). All these members of the Coastal Strand community are widespread in this community, and most of them occur only in this distinctive plant community. The Coastal Strand plant community occurs in loose sand above the high tide line on Pacific coast beaches along the entire length of California, and extends both south and north of the state's borders. Some of the species in this community are very widespread outside California. For example, Beach Morning Glory is found all along the Pacific coast of North America and in South America and the Old World as well.

74

Herbaceous

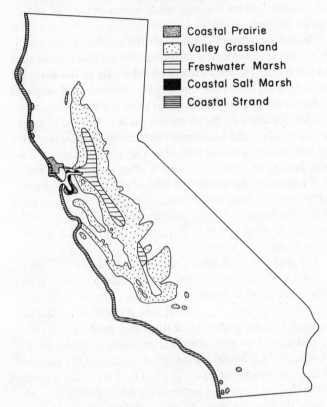

Coastal Prairie

Valley Grassland

Freshwater Marsh

Coastal Salt Marsh

Coastal Strand

Map. 2. Approximate distribution of herbaceous plant communities in California.

The climate of the area occupied by the Coastal Strand plant community in California is variable. Rainfall varies from an average annual total of about 15 inches (38 cm) to a total in excess of 70 inches (178 cm). In many parts of the coast, there are extensive fogs during the summer and the winter. The growing season along the immediate coast is very long and even in extreme northern California exceeds 350 days. Because of the tempering effect of the adjacent ocean, diurnal as well as seasonal temperature fluctuations are relatively small. Summers are cool and winters are relatively warm.

Despite the relatively benign climatic conditions outlined above, the environment of the Coastal Strand plant community is a harsh one. This may explain why there are so few plant species that occur in this community. The community is subjected to strong winds during much of the year, and these carry salts that are deposited on the plants and the soil. Because of these salt-laden winds and occasional high tides during winter storms, the sand occupied by the plants has a high concentration of sea salts. Besides being salty, the sand is often very unstable and blown about by the winds. The level of plant nutrients is generally very low; from a nutritional standpoint these sands are infertile. Furthermore, during summer months the surface of the sand may become extremely hot, to the extent that it is very uncomfortable to the human touch.

Although the plant species that occur in the Coastal Strand are taxonomically unrelated, many of them share a number of similar adaptive characteristics. For example, many of the plants in this community are prostrate and have creeping stems that hug the sand. In some species, these stems may produce roots at the nodes and eventually form a large colony derived from a single individual. Sexual reproduction of some of the perennial species may be relatively rare, perhaps because of the great difficulties that seedlings have in becoming established in the continually shifting and generally inhospitable sands. Also, many Coastal Strand plants have grayish foliage; this is probably an adaptation to the frequent extreme daytime heat to which these plants are subjected and undoubtedly serves to reflect heat from the plant and thus reduce the temperature of internal tissues. Plants in this community are frequently succulent and may have sufficiently high salt concentrations in their tissues to be detectible to the human taste. Possibly this succulence is an adaptation to the occasional dry periods to which strand plants are subjected, and it may also enable these plants to take in water whose salt content would dehydrate most non-succulent plants of

76

other communities. Needless to say, Coastal Strand plants are notably resistant to salt and to wind.

Coastal Prairie (Plate 5B, Map 2)

Immediately inland from the Coastal Strand plant community there are extensive grasslands known as Coastal Prairie. This plant community occurs sporadically along the northern California coast from the Oregon border to the San Francisco Bay area. At one time, some of the hills behind Oakland and Berkeley were occupied by Coastal Prairie as well as by patches of Valley Grassland. Some of the "balds" on the northern California coastal hills — which may be several miles from the ocean — represent the Coastal Prairie plant community. Extensive areas along the coast of northern California have been cleared to encourage the growth of grasses for grazing animals; such areas would seem to belong to the Coastal Prairie plant community. Because of their unnatural origin, however, and their component of introduced plant species, these artificial grasslands should be excluded from a definition of the Coastal Prairie plant community. Coastal Prairie was originally covered with a number of native perennial bunch grasses mixed with several other herbaceous plants. Shrubs and trees are missing. The soils of Coastal Prairie areas are typical prairie soils similar to those found in the grasslands of the American Midwest, indicating that these coastal areas have been occupied by prairie for hundreds of years or more and are not of recent origin. For this reason, Coastal Prairie can be considered to be a climax plant community in California.

Typical plants of the Coastal Prairie are perennial grasses that belong to genera such as *Festuca, Danthonia, Calamagrostis, Deschampsia,* and others. Bracken Fern (*Pteridium aquilinum,* Pteridaceae) also is a common inhabitant of this area. Monocots such as *Brodiaea pulchella* (Amaryllidaceae), Douglas Iris (*Iris douglasiana,* Iridaceae — in blue and white forms), Blue-eyed Grass (*Sisyrinchium bellum,* Iridaceae), and Yellow Butterfly Tulip (*Calochortus luteus,* Liliaceae) add a colorful display of flowers in late spring months. Also present are California Buttercup (*Ranunculus californicus,* Ranunculaceae), two lupines

77

(*Lupinus variicolor* and *L. formosus*, Leguminosae), the peculiar prostrate Yellow Mats (*Sanicula arctopoides*, Umbelliferae), Golden Aster (*Chrysopsis villosa* var. *bolanderi*), and a few other members of the Compositae.

Because Coastal Prairie areas are naturally treeless and because they occur in temperate and relatively well-watered areas of California, the majority of the area occupied by the Coastal Prairie plant community has been subjected to grazing by sheep and cattle since the settlement of California by agriculturalists. Many of the areas along the coast that are being developed for summer housing tracts also occur in areas occupied by the Coastal Prairie plant community.

Coastal Salt Marsh (Plate 5C, D; Map 2)

Coastal Salt Marsh is a plant community which exists under some of the same climatic conditions as the Coastal Strand community although there are some striking differences in the ecology of the two communities. As the name implies, the Coastal Salt Marsh community occurs along the Pacific coast of California (and adjacent areas), although it is a much less frequently encountered community than is the Coastal Strand community. The Coastal Salt Marsh community is found in estuaries, bays, and other areas that are protected from the wave action and strong winds of the open coast. This community is widely scattered on the Pacific coast; in California it occurs at the edges of Humboldt Bay, San Francisco Bay, Tomales Bay, Morro Bay, the vicinity of Santa Barbara, the San Diego area, and a few other areas.

The soil is generally very wet and in some areas is periodically inundated with salt water by tidal action. As a consequence of the water-saturated soils (which are often heavy clays), the roots of salt-marsh plants occur in a soil with a very low oxygen concentration. Because of the salinity of the salt-marsh soils and of the water that reaches the plants growing on these soils, salt-marsh plants are all *halophytes*, that is, plants that live in saline soils. The halophytes of the Coastal Salt Marsh community are mostly low perennial herbs with fleshy stems and leaves. Often, the leaves are very much reduced. Salt-marsh plants are

78

frequently rhizomatous and, like strand plants, may reproduce vegetatively. The flowers of most salt-marsh plant species in California are inconspicuous, although this characteristic is not obviously related to any ecological peculiarities of the habitat.

Typical Coastal Salt Marsh species in California are members of the peculiar glasswort or pickleweed genus *Salicornia* (Chenopodiaceae, Plate 5D); sea blite or seep weed (*Suaeda* spp., Chenopodiaceae); Cord Grass (*Spartina foliosa*, Gramineae); Salt Grass (*Distichlis spicata*, Gramineae); Sea Lavender or Marsh Rosemary (*Limonium californicum*, Plumbaginaceae); Frankenia (*Frankenia* spp., Frankeniaceae); and arrow grass (*Triglochin* spp., Juncaginaceae). There are rather few plant species represented in the Coastal Salt Marsh; like the Coastal Strand plant community it is relatively impoverished in number of species and also relatively monotonous because its component species tend to occur widely in salt marsh areas along the coast.

Pickleweed is, as its name suggests, used for making pickles. Its salty, succulent, jointed stems are essentially tasteless, when fresh, but when impregnated with vinegar, sugar, and spices, they make an acceptable and inexpensive substitute for cucumber pickles. Although few people in California use *Salicornia* for making pickles, this practice is still fairly common in northern Europe. Because of the high content of sodium (from salt) in the tissues of the plants, in times past large quantities were burned for soda ash which in turn was used for making glass. Thus, the two common names for *Salicornia*, pickleweed and glasswort, are derived from actual uses of the genus. During May and continuing into summer months, the plants of coastal salt marshes become infested with the colorful bright-orange strands of the dodder, *Cuscuta salina* (Cuscutaceae), a parasitic relative of the morning glory (Plate 5D).

Some portions of the Coastal Salt Marsh plant community are subjected to inundation due to tidal action; other areas are only occasionally inundated, although the soil may be saturated with salt water. Some plant species characteristic of this plant community are able to tolerate immersion in salt water and others are not. As a result, there is frequently a zonation pattern of plants in a Coastal Salt Marsh that is related to the tolerance of

79

the plants to inundation. For example, Cord Grass tolerates inundation but Salt Grass is intolerant of frequent inundations. Frankenia does not occur in areas inundated by tidal action. It is not surprising that many genera of plants that occur in Coastal Salt Marsh also reappear in saline areas of the desert. Salt Grass and pickleweed have related species in desert regions.

The area covered by the Coastal Salt Marsh plant community is relatively small, and in California is mostly in regions where there are high concentrations of human population. In the San Francisco Bay area, hundreds of acres of salt marsh have been lost by bay-fill projects of various kinds. Likewise, in southern California a number of salt marshes have been reduced in size (or perhaps lost) due to draining and other ecological disturbances. If pollution of San Francisco Bay continues to increase, the altered biotic conditions at the periphery of the bay may have a deleterious influence on the Coastal Salt Marsh community, although these effects have yet to be demonstrated. From a zoological standpoint, this community is valuable because it provides feeding and nesting areas for a large number of resident or migratory water birds. Most efforts at conserving the community have been directed toward the zoological rather than botanical interest of the community, but both merit the attention of conservationists because of their unique qualities.

Northern Coastal Scrub (Plate 6A; Map 3)

Continuing in an eastward transect across the coastal portion of northern California, immediately inland from the Coastal Prairie plant community there is another plant community that is dominated by a maritime climate but which differs conspicuously from the Coastal Prairie plant community in that shrubs are present. This plant community is the Northern Coastal Scrub, which occupies a narrow and discontinuous strip of land running from the Oregon border south to Santa Cruz County, reappearing briefly south of Monterey Bay from Pacific Grove to Point Sur. It lies between the Coastal Prairie and the North Coastal Forest or the Closed-Cone Pine Forest discussed below. The Northern Coastal Scrub is characterized by the presence of

Scrublands

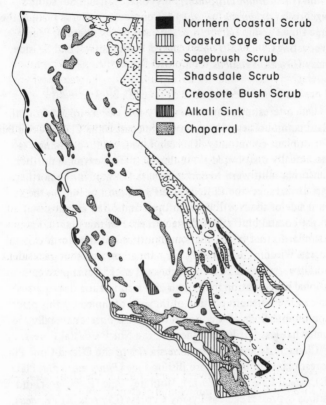

Northern Coastal Scrub
Coastal Sage Scrub
Sagebrush Scrub
Shadsdale Scrub
Creosote Bush Scrub
Alkali Sink
Chaparral

Map 3. Approximate distribution of scrubland plant communities in California.

low shrubs intermixed with grassy meadows. These small meadows contain plant species that are characteristic of the Coastal Prairie. Shrubs present in the Northern Coastal Scrub are Coyote Brush, *Baccharis pilularis* var. *consanguinea* of the Compositae (a prostrate coastal form of this species is becoming increasingly popular as a ground cover plant for planting in dry, sterile, ground); Seaside Woolly Sunflower (*Eriophyllum staechadifolium*, Compositae); Salal (*Gaultheria shallon*, Ericaceae); Coastal Eriogonum (*Eriogonum latifolium*, Polygonaceae); and Suksdorf's Sagebrush (*Artemisia suksdorfii*, Compositae). Herbs include the large Cow Parsnip (*Heracleum lanatum*, Umbelliferae); Pearly Everlasting (*Anaphalis margaritacea*, Compositae); and Seaside Daisy (*Erigeron glaucus*, Compositae).

Closed-Cone Pine Forest (Plate 6B, D; Map 4)

One interesting plant community that occurs inland from the plant communities of the immediate coast is the Closed-Cone Pine Forest plant community. This coniferous plant community is sporadically distributed along the coast from extreme northern California southward to Santa Barbara County. In the northern part of its range, the Closed-Cone Pine Forest occurs on the seaward side of the North Coastal forest and often extends well out on the coastal bluffs. The climatic regime of this plant community is similar to that of the coastward portions of the North Coastal Forest. Winters and summers are temperate, frosts are rare, and rainfall varies from 20 to 60 inches (51 to 152 cm) per year. Additional precipitation may occur as a result of fog drip, particularly in the northern portion of the area occupied by this plant community. Soils of the Closed-Cone Pine Forest generally are somewhat less fertile than those of the North Coastal Forest.

Characteristic coniferous tree species of the Closed-Cone Pine Forest plant community are Bishop Pine (*Pinus muricata*, Plate 6B), Beach Pine (*P. contorta*), Monterey Pine (*P. radiata*), and various cypresses such as Pygmy Cypress (*Cupressus pygmaea*), Gowen Cypress (*C. goveniana*), and Monterey Cypress (*C. macrocarpa*). Bishop Pine, Beach Pine, and Pygmy Cypress commonly occur in this community north of San Francisco Bay, and Monterey Pine, Monterey Cypress, and Gowen Cypress occur only south of the Bay. The term "closed-cone" comes from the fact that the cones of the pines in this forest do not open at maturity but remain closed for several years after maturation, and then gradually open and disperse seeds.

Although generally not a member of the Closed-Cone Pine Forest plant community, another closed-cone pine in California

Woodlands

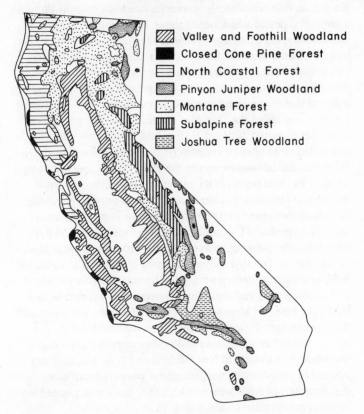

Valley and Foothill Woodland
Closed Cone Pine Forest
North Coastal Forest
Pinyon Juniper Woodland
Montane Forest
Subalpine Forest
Joshua Tree Woodland

Map 4. Approximate distribution of woodland plant communities in California.

is Knobcone Pine (*Pinus attenuata*, Plate 6C), a species which mostly occurs in inland localities and is particularly common on burned-over lands. The cones of Knobcone Pine open during the heat of a fire; thus, this species has a built-in mechanism that allows it to re-seed itself after a fire.

The closed-cone pines have an extensive fossil history which provides an insight into their evolutionary background. The ecological distinctiveness of the coastal species compared with Knobcone Pine is associated with the fact that the fossil record of

Knobcone Pine is essentially a separate one back into the Pliocene, a geological period which began about 10 million years ago. In other words, the genealogy of Knobcone Pine indicates that it has had an ancestry separate from that of its close relatives for several million years. In contrast, the coastal closed-cone pines were derived from a common ancestor that prevailed in the Pleistocene, a period that began about one million years ago. (The Pleistocene is sometimes called the "Ice Age" in allusion to the great glaciations that occurred during that time.) All closed-cone pines are connected to a common ancestor (*P. masonii*) in the Pliocene. *Pinus masonii* became extinct in the early Pleistocene, although some of its traits persist in its modern descendants. The fossil record also indicates that Monterey Pine was once considerably more widespread in California than it is at present. Like some other tree species in California, Monterey Pine has had its day, evolutionarily speaking, and is on its road to natural extinction. As a matter of fact, the only closed-cone pine which seems to be holding its own in contemporary California is Knobcone Pine. It is an interesting and somewhat puzzling contrast that both Monterey Pine and Monterey Cypress — rare and remnant species that are ecologically unsuccessful in their natural habitat — are widely planted as windbreaks and timber trees in many areas of the world, particularly in New Zealand and Australia, and that in some of these southern regions these trees are invasive weeds that have spread from areas in which they have been planted into adjacent native plant communities.

A particularly interesting local phase of the Closed-Cone Pine Forest occurs in the Mendocino White Plains area of Mendocino County in northern California. This phase is locally known as the Pygmy Forest and occurs on the highly acid, sterile soils a few miles inland from the immediate coast. The soil acidity and the sterility are coupled with the fact that these soils are underlain with an impervious hardpan and are waterlogged for much of the winter rainy season. This combination of unfavorable ecological characteristics results in the presence of an assemblage of dwarfed trees, or natural bonsais, hence the name "Pygmy Forest" (Plate 6D). Trees of Pygmy Cypress (*Cupressus pygmaea*) and Bolander Pine (*Pinus contorta* var. *bolanderi*) that occur on this soil may reach a mature height of only one or two feet (0.3 to 0.6 m) in 50 or 60 years. These dwarfed trees occur with a few other plant species, some of which are restricted to the White Plains soils (e.g., the Fort Bragg Manzanita, *Arctostaphylos nummularia*). However, if Pygmy Cypress or other tree species that occur on these soils break through the hardpan or become established on more normal soils, they will grow to the usual tree-like proportions that one expects. In this case, therefore, the dwarfing is induced by the unusual soil characteristics and is not a part of the genetic makeup of the plants that occur in the area.

North Coastal Forest (Plates 7; 8A, B; Map 4)

The greatest difference between the Munz and Keck treatment of California plant communities and that presented here is in the interpretation of the North Coastal Forest. Munz and Keck break up this collective plant community into several plant communities: the North Coastal Coniferous Forest, Redwood Forest, Douglas Fir Forest, and Mixed Evergreen Forest. Arguments can be offered pro and con the conservative treatment used here, but it is simpler to treat this complex and variable assemblage of woody plants as a single plant community. The North Coastal Forest occurs over much of the North Coast Ranges, in the Sis-kiyou-Klamath mountains, and in the Santa Cruz mountains. It continues northward along the Pacific coast into southern Alaska; patches of forest referable to this plant community occur in the vicinity of Mount Shasta as well, although in this last area most of the typical conifers are absent and the plant species present are those characteristic of Munz and Keck's Mixed Evergreen Forest.

Over most of its area of distribution, the North Coastal Forest is dominated by one or more species of coniferous trees. These include majestic species such as Coast Redwood (*Sequoia semper-virens*, Plate 7A), Douglas Fir (*Pseudotsuga menziesii*, Plate 7A), Western Hemlock (*Tsuga heterophylla*), Lowland Fir (*Abies grandis*), Sitka Spruce (*Picea sitchensis*), Canoe or Western Red Cedar (*Thuja plicata*), and Port Orford Cedar (*Chamaecyparis lawsoniana*). Also present are the Giant Chinquapin (*Chrysolepis*

chrysophylla), Tanbark Oak (*Lithocarpus densiflora*), and Bigleaf
Maple (*Acer macrophyllum*), which are hardwood trees associated
with the conifers. Understory species consist of small trees or
shrubs such as Vine Maple (*Acer circinatum*, Plate 8B), Evergreen
or California Huckleberry (*Vaccinium ovatum*), Salal (*Gaultheria
shallon*, Plate 7D), California Rose Bay (*Rhododendron macrophyllum*), Thimble Berry (*Rubus parviflorus*), Sword Fern (*Polystichum munitum*), and Redwood Sorrel (*Oxalis oregana*).

Climatically, the area occupied by the North Coastal Forest is
variable. Average rainfall varies from over 100 inches (254 cm)
in redwood areas to as low as 25 inches (64 cm) in inland areas
occupied by forests of Douglas Fir or the phase of the North
Coastal Forest that Munz and Keck designated as the Mixed Evergreen Forest. Where this plant community occurs on the immediate coast it is in a cool area dominated by coastal fog during much
of the summer; inland, summer temperatures become increasingly
higher and the winter temperatures increasingly lower. These
gradients in moisture and rainfall are reflected in differences in
the species composition of the North Coastal Forest throughout
its area. In the wetter portions of its range, this forest is dominated by the Coast Redwood, Sitka Spruce, Western Hemlock,
Canoe Cedar, or Port Orford Cedar. In somewhat drier areas,
Douglas Fir is the dominant tree species. In even drier, warmer
areas — usually on the inland slopes of the North Coast Range —
conifers give way to various broad-leaved trees such as Kellogg
Oak (*Quercus kelloggii*), Canyon Oak (*Q. chrysolepis*), Tan Oak
(*Lithocarpus densiflora*), Madrone (*Arbutus menziesii*), California
Bay (*Umbellularia californica*), California Hazelnut (*Corylus
cornuta* var. *californica*), Mountain Dogwood (*Cornus nuttallii*).
and various species of the shrubby California lilac (such as *Ceanothus parryi* and *C. thyrsiflorus*). As the average rainfall decreases
even more, the North Coastal Forest may be replaced by Valley
and Foothill Woodland, Valley Grassland, or in some areas,
Chaparral.

Because of its floristic complexity, it is difficult to make
many meaningful generalizations about the North Coastal Forest.
However, a few species deserve special comment. Perhaps the
best known phase of the North Coastal Forest is that segment of

it dominated by Coast Redwood, *Sequoia sempervirens*. This tree occurs in the coastal fog belt of northern California and extends into southern Oregon. The Coast Redwood has rather few other plant species growing with it. Like various species of closed-cone pines, Coast Redwood had a much wider distribution in the geological past than it does at present and should be considered a species that is on its way out in an evolutionary sense.

Fossil History of Coast Redwood

The Coast Redwood has an interesting fossil history that was not fully understood until quite recently. In publications dealing with the fossil plants of the northern hemisphere, one encounters statements regarding the wide distribution of "redwoods" in the geological past. These redwoods in the fossil record are widely distributed around the northern hemisphere. For many years these fossils were all referred to the genus *Sequoia*, and most of them were considered to represent very close relatives of the modern Coast Redwood. In the early 1940's, however, a Japanese botanist noted that the arrangement of leaves and the cone structure of some of these fossil remains were different from those of Coast Redwood. He named the fossils with opposite leaves and stalked cones as a new genus, *Metasequoia*. Re-examination of the fossil redwoods indicated that *Metasequoia* occurred in North America, Greenland, and Asia, although these fossils had previously been referred to *Sequoia.* Fossil *Sequoia* is known from western Europe, Greenland, portions of Asia, and also from North America.

In the mid 1940's a Chinese forester reported the occurrence of an unusual and previously unknown conifer growing in the province of Szechuan, central China. Subsequent examination of branches of this tree revealed that it was a living plant of *Metasequoia!* Thus, the genus was known to science first as a fossil and then was later discovered to be living. The living specimens were named *M. glyptostroboides* (the specific name after *Glyptostrobus*, a genus of Taxodiaceae related to the redwoods). A few small groves of *Metasequoia* were later located in the vicinity of the original "Discovery Tree". The seed that was collected and sent to the United States all originated from this tree, which had survived

87

because it was venerated and protected by local people. *Metase-quoia*, or the Dawn Redwood, is now widely planted as an ornamental tree and is able to tolerate a wide variety of ecological conditions. Unlike Coast Redwood, *Metasequoia* is deciduous (like its relatives in the genus *Taxodium*, the bald cypresses).

The fossil history of Coast Redwood goes back several million years and indicates that this species and its immediate ancestors were once scattered over much of the northern hemisphere. Climatic changes in the relatively recent geological past have resulted in the gradual extinction of *Sequoia* over much of its former range, with the result that *S. semperivirens* is now restricted to the outer Coast Ranges from extreme southern Oregon southward to Santa Cruz County (more or less continuously), with a few outlying groves in Monterey County, California. The rainfall in this coastal strip is variable, ranging from a high of over 100 inches (254 cm) annually in the north to a low of 35 inches (89 cm) in the south. However, since Coast Redwood is generally restricted to the coastal fog zone, transpiration during the dry summers is much reduced by the effect of the fog. In addition, as much as 10 inches (25 cm) of added precipitation per year has been recorded as a result of fog drip under trees, with the result that Coast Redwood is able to exert a favorable influence over its immediate environment by adding to the effective precipitation.

The phase of North Coastal Forest occupied by Coast Redwood is esthetically very pleasing but rather monotonous botanically. There are rather few plant species associated with Coast Redwood; the forest floor under these gigantic trees is remarkably uniform in terms of the few herbs that grow under the trees. One finds Sword Fern (*Polystichum munitum*), Redwood Sorrel (*Oxalis oregana*), Inside-out Flower (*Vancouveria parviflora*), and the slightly woody, trailing saxifrage, Yerba de Selva (*Whipplea modesta*). Characteristic shrubs include California or Evergreen Huckleberry (*Vaccinium ovatum*), California Rose Bay (*Rhododendron macrophyllum*), Salal (*Gaultheria shallon*), and Wax Myrtle (*Myrica californica*). In some areas, Coast Redwood co-exists with California Bay (*Umbellularia californica*), Douglas Fir (*Pseudotsuga menziesii*) and Tan Oak (*Lithocarpus densi-*

88

flora). However, these trees are not very successful competitors with Coast Redwood, for reasons that will be discussed below.

The fossil history of Coast Redwood suggests that it has grown with its present major plant associates or their ancestors for some time. Some years ago, Ralph Chaney, of the Department of Paleontology at the University of California, Berkeley, examined the fossil flora from deposits near Bridge Creek in the John Day River basin of north-central Oregon. This fossil deposit dates from Miocene times and represents an assemblage of fossil plants that grew together in that part of Oregon between fifteen and twenty million years ago. The present xerophytic vegetation of the Bridge Creek area is a very different one from that present in Miocene times. Chaney identified almost 21,000 individual fossil remains from the Bridge Creek deposit. Fragments of Coast Redwood or its Miocene counterpart were common in these remains. Then Chaney scooped up and catalogued the plant fragments that had been deposited in recent times in the bed of a stream running through Muir Woods in Marin County, California, an area occupied by a modern Coast Redwood forest. The following chart compares the composition of the Miocene Redwood forest in Oregon and the modern Coast Redwood forest in Marin County:

MUIR WOODS	FREQUENCY (In Percent)	BRIDGE CREEK	FREQUENCY (In Percent)
Sequoia sempervirens	39	*S. langsdorfii*	15
Alnus oregona	27	*A. carpinoides*	54
Lithocarpus densiflora	5	*Quercus consimilis*	9
Umbellularia californica	13	*Umbellularia* sp.	9
	84		87

The chart requires some explanation. The column on the left gives the name of the four tree species that are most commonly encountered in the bed of the Muir Woods stream as leaf, twig, or cone fragments. The proportion of the plant remains that is made up by each of them is also given. That is, 39 percent of the fragments found in the Muir Woods stream are Coast Redwood twigs, leaves, and cones. The column on the right gives the figures for the representation of the four most common woody species encountered in the Miocene Bridge Creek flora. The

89

binomials given in this right-hand column may be unfamiliar ones, but they are the names that are given to the fossil counterparts of the species listed on the left. That is, *Sequoia langsdorfii* is the Bridge Creek counterpart of *S. sempervirens; Quercus consimilis* is the counterpart of the modern *Lithocarpus densiflora*. The figures for each of the four species are somewhat different between the Miocene representatives and the modern Redwood Forest counterparts, although the total figures for the four species at each time period are similar (84 percent vs. 87 percent). Since we are dealing with the representation of plants that grew in two widely separated areas at time periods that are separated by fifteen or twenty million years, it is remarkable that there is so strong a similarity between the species composition of the Miocene redwood forest and the modern one. It would seem that Coast Redwood and its woody associates have had an amazingly long and relatively consistent association over a period of many millions of years. It is probable that this is also the case for a number of other plant communities in California. Some plant communities of the state — such as the Closed-Cone Pine Forest or segments of the North Coastal Forest (as exemplified by the Coast Redwood forest) — are very "old" and conservative plant communities. Other California plant communities, to be discussed later, are probably relatively recent newcomers that have evolved within the past million years or so.

Ecological Characteristics of Coast Redwood

There are a few other tree species that occur with Coast Redwood, although perhaps with only a modest degree of success. Some of the ecological requirements of Coast Redwood have been mentioned above. Obviously, areas in California with a "redwood climate" generally support a Coast Redwood forest. But what is it about Coast Redwood that allows it to be so successful in these areas, almost to the exclusion of other species of herbs, shrubs, or trees? Coast Redwood casts considerable shade and its roots pervade the ground under the trees. In addition, the ground under the trees is covered with a layer of needles, branches, and other plant debris that may be several centimeters deep. This combination of shade, root competition, and a deep organic layer on the soil surface probably is effective in reducing the number of species of plants that can thrive under Coast Redwood trees.

Study of the soils and past history of Coast Redwood forests indicates that these forests have been subjected to periodic burning and also to periods of heavy silting as a result of floods (such as the periodic floods that have devastated the redwood region in recent years). These two ecological factors are probably very important in eliminating potential tree competitors of Coast Redwood. Coast Redwood is fairly tolerant of fires, and even young plants can produce new shoots from the roots or lower trunks if the upper portions of the tree are completely destroyed (Plate 7B). Some of the potential tree competitors of Coast Redwood are not fire tolerant, however, and are completely destroyed by these periodic fires. Furthermore, Coast Redwood can tolerate silting by floods since the species can produce new surface-feeding roots either by developing new roots from the old ones after they have been buried under silt, or by the production of completely new roots systems from the trunks of the tree just below the surface of the silt deposit. In contrast, some of the potential tree competitors of Coast Redwood are intolerant of silting and die very soon after their roots have been covered by the layer of silt deposited during floods.

The potential tree competitors of Coast Redwood over much of its area of distribution are Tan Oak, Douglas Fir, Grand Fir, and California Bay. None of these species is very tolerant of either silting or fire. Even if individuals of these species become established in an undisturbed Coast Redwood forest, they are at a competitive disadvantage compared with Coast Redwood. For example, California Bay is a relatively slow grower and rather intolerant of shade. Consequently, one rarely encounters a full-grown or fully vigorous California Bay tree in a well-established Coast Redwood forest. Douglas Fir can compete successfully with Coast Redwood if both species start out simultaneously in an area as seedlings, but Douglas Fir will eventually disappear from the area due to the shorter life span of this species and because its seedlings cannot become established in the dense shade of Coast Redwood. Similar explanations can be offered for other

91

tree species that occur in the redwood region but which generally do not occur in a vigorous state in association with Coast Redwood. Because of the intolerance of Coast Redwood to prolonged drought, shallow soils, and hot climates, this species is replaced on the eastward side of its range (in many areas) by Douglas Fir, a species which tolerates these conditions rather well. Thus, the mosaic effect of local tree dominants in the North Coastal Forest can be explained by the individual ecological characteristics of the local dominants as well as of their potential competitors.

Chaparral (Plate 8C, D; Map 3)

Another of the plant communities that occur along our transect is Chaparral. Chaparral is one of the most characteristic plant communities of California, and occurs only in the California Floristic Province. Chaparral is a broad-leaved *sclerophyll* type of vegetation. Sclerophyll means "hard-leaved", in reference to the hard, stiff, thick, heavily cutinized, and generally evergreen nature of the leaves of chaparral shrubs. This type of leaf is characteristic of many xerophytic shrubs. The shrubs that dominate Chaparral are generally rather low, the average being between 3 and 6 feet (0.9 and 1.8 m) tall, although occasional individuals may reach up to 10 feet (3 m). Chaparral is dense, often impenetrable, and notably deficient in trees and herbs. Indeed, the ground underneath or among chaparral shrubs is often completely devoid of herbaceous plant species. This may be due in part to shading or to competition from roots of chaparral shrubs for water, although much of this phenomenon is probably also due to allelopathy, since some chaparral shrubs (e.g., Chamise, *Adenostoma fasciculatum*) are known to exhibit allelopathy.

The word chaparral is of Spanish origin. In Spain, "chaparro" refers to a scrub oak. The suffix "-al" means "a place of". Thus, chaparral is "a place of scrub oak". In California, the term Chaparral came to be applied to a specific type of plant community consisting of a dense growth of evergreen, hard-leaved shrubs, although taxonomically these shrubs are mostly not oaks. Characteristic species of Chaparral are Chamise or Greasewood (*Adenostoma fasciculatum,* Rosaceae), California Holly or Toyon (*Heteromeles arbutifolia,* Rosaceae), Holly-leaf Cherry *(Prunus*

92

ilicifolia, Rosaceae), Mountain Mahogany (*Cercocarpus betuloides,* Rosaceae), and various species of manzanita (*Arctostaphylos* spp., Ericaceae) and California lilac (*Ceanothus* spp., Rhamnaceae). Also present are Scrub Oak (*Quercus dumosa*) and Poison Oak (*Rhus diversiloba,* Anacardiaceae). In southern California, additional species in Chaparral are Spanish Bayonet (*Yucca whipplei,* Agavaceae), Laurel Sumac (*Rhus laurina,* Anacardiaceae), and Sugar Bush (*Rhus ovata*).

The foregoing are the shrub species that are characteristic of Chaparral. It should be pointed out, however, that numerous other plant species occur in Chaparral. A tally of plants that commonly occur in Chaparral in California indicates that nearly 900 species of vascular plants occur in this community. About 240 of these are woody plants, most of which are shrubs; well over 300 are annual or biennial herbs; an equal number of perennial herb species is present. The largest plant families present are the Compositae (sunflowers) with over 90 species in Chaparral, Scrophulariaceae (figworts) with nearly 70 species, and Gramineae (grasses), Leguminosae (legumes), Hydrophyllaceae (waterleafs), Labiatae (mints), Ericaceae (heaths), and Liliaceae (lilies) with between 30 and 40 species of each. Thus, the list of characteristic species in any plant community is a gross oversimplification of what one can expect to find there. The species that have been chosen as typical, however, generally are widely distributed, common, and conspicuous in these plant communities, and because of these traits they are listed in preference to less conspicuous species, less widely distributed ones, or ones with less of an ecological fidelity.

Chaparral occurs in areas with wet, mild winters and long, dry, hot summers. Average annual rainfall ranges from about 15 inches to 25 inches (38 to 64 cm). However, summer rainfall accounts for less than 20 percent of the annual total rainfall. Soils occupied by Chaparral are often gravelly or sandy, shallow, and have a low water-holding capacity. The distribution of Chaparral in California is a "spotty" one which shows no clear geographic coherence. A probable explanation for this is that Chaparral often occurs in an area having a climate that would lead one to expect a woodland community, but because of the local unfavorable soil situation, forest communities are absent and are replaced by a shrub community.

The flowering and growth behavior of chaparral shrubs merits some comment. Some shrubs, such as some manzanitas, typically flower in mid-winter (sometimes in December) or very early spring and produce vegetative growth after flowering. At the termination of the vegetative growth, flower buds are produced which remain dormant throughout the succeeding summer and autumn drought, although they burst into flower very quickly in winter. In contrast, some other species (such as Chamise) produce vegetative growth in late winter or early spring and then flower in June. The flower buds of these species are produced at the termination of growth as they are in winter-flowering species, except that there is no bud dormancy; flowering occurs immediately after development of flower buds. One result of this phenomenon is that most chaparral shrubs have a similar growing period, but flowering in the community may extend over a period of six months or longer, since various species have different flowering periods. The majority of shrubby species in Chaparral flower in April, however, when soil moisture conditions are optimal and the air and soil temperatures also are optimal.

By definition, Chaparral is a plant community in which the dominant shrub species are evergreen. However, most of the chaparral species exhibit a leaf drop of old leaves in early summer, during or at the end of the vegetative growth of new shoots. Because of this timing, the plants retain some leaves the year round. After growth and leaf drop occur, the shrubs become dormant. When the subsequent wet season arrives in late autumn, the chaparral shrubs are ready to go; they do not have to produce a fresh crop of leaves as do deciduous species. In this respect, at least, chaparral shrubs show an interesting adaptation to the climatic regime under which they exist, an adaptation that enables them to take advantage of the limited period of rainfall.

Fire and Chaparral

Chaparral is subject to frequent burning over much of its area of distribution and, indeed, in much of its range it is a subclimax plant community which is maintained because of periodic fires.

Many chaparral species show an interesting ecological adaptation to repeated burning. This adaptation is *crown-sprouting,* which refers to the fact that although the above-ground portions of a chaparral shrub may be destroyed by a hot brush fire, the individual produces numerous new shoots that develop from a large burl that terminates the root system at or below the soil surface (Plate 4B). As a result of this trait, crown-sprouting chaparral shrubs are able to re-establish themselves immediately after a fire and do not go through a successional re-establishment procedure via seedlings. However, some chaparral shrubs are not crown-sprouters, and these species are killed by hot fires and must produce seedlings in order to become re-established on a site. The advantage of crown-sprouting is that it eliminates the uncertainties of seedling establishment, which in arid climates may be considerable.

Chaparral is prone to burning because the shrubs are dense, close together, and have rather dry evergreen leaves. Once a fire gets started in Chaparral, it may spread rapidly and extensively. The heat of chaparral fires is often intense. Temperatures of 1200°F (649°C) have been recorded at the surface of the soil in burning Chaparral; one and one-half inches (3.8 cm) below the soil surface the temperature may reach well over 300°F (149°C). The biotic effects of chaparral fires are several. One of these is the heating of the soil; another is that the fires remove the vegetation and thus allow more light to reach the soil surface. The improved light conditions, plus the absence of root competition and the vaporization of allelopathic compounds of shrubs, allow the establishment of a lush herbaceous flora the first season after chaparral fires. Although herbaceous annual plants are notably uncommon under or near chaparral shrubs, their immediate and abundant appearance after a fire indicates that their seeds are present in the ground.

There are a few plant species in California that are found only after fires have swept through an area. These include Whispering Bells (*Emmenanthe penduliflora,* Hydrophyllaceae) and one of the native snapdragons, *Antirrhinum cornutum* (Scrophulariaceae), both annuals. Other species such as the poppy (*Papaver californicum,* Papaveraceae), *Phacelia brachyloba* (Hydrophyl-

laceae, Plate 4D), and Golden Corydalis (*Corydalis aurea*, Fumari-
aceae) are common on burned areas, although they may also
appear in areas that have been ecologically disturbed by other
agencies. The fire annuals listed above persist for many years as
seed in Chaparral; indeed, the viability of their seed may last as
long as a hundred years or more. In areas that have not been
burned for sixty years, study of the soil has revealed the presence
of viable seeds of these plant species, most of these seeds produced by the previous generation of plants that grew on the site
when it last burned. Seeds of fire annuals are obviously tolerant
of the very high soil temperatures that result from chaparral fires.
Also, many of these species require a high-temperature shock in
order to germinate, and thus have a built-in mechanism which
informs the seeds that a fire has occurred. In the autumn after
the fire, germination of seeds occurs after the first heavy rains
fall; in the subsequent spring, the former chaparral site is covered
with thousands of individuals of these fire annuals, the offspring
of parents that occupied the site up to a century before. These
fire annuals are accompanied by various other annual plant
species that are not necessarily typical of burned areas. Some
herbaceous perennials also appear in profusion on burned lands.
However, all these plants are also accompanied by seedlings of
chaparral species and by the rapidly growing shoots of the crown-
sprouting shrubs. As a result, within a few years the chaparral
shrubs have re-occupied the site and the annuals disappear — until
another fire occurs.

Most chaparral areas that have been studied show a recent history of fire. In areas where Chaparral is not a climax plant community, it is likely to be replaced successionally by one of the
various woodland communities present in the vicinity. Seedlings
of trees that become established among chaparral shrubs eventually increase in size and abundance, and the chaparral shrubs
correspondingly begin to decline. However, there are areas where
Chaparral is maintained because the local soil conditions are un-
suitable for the establishment of tree species. In these circumstances, Chaparral is a climax plant community.

Valley and Foothill Woodland (Plate 9A-C; Map 4)

Large areas of the valleys and eastern slopes of the North and South Coast Ranges, the valleys of interior southern California, and the western foothills of the Sierra Nevada are occupied by the Valley and Foothill Woodland plant community. This plant community occurs at elevations ranging from 300 or more feet (100 m or more) above sea level to as high as 5,000 feet (1,524 m) in southern California. It is characterized by scattered trees with an undergrowth that may consist almost exclusively of herbaceous plants, especially grasses, and scattered low shrubs; in some areas (the phase characterized by Munz and Keck as Foothill Woodland) the trees may be rather dense, with scattered shrubs underneath them. Variations in appearance of this plant community depend to some extent upon its location as well as upon its species composition.

In some areas, the Valley and Foothill Woodland is dominated chiefly by oaks, such as Garry Oak (*Quercus garryana*), Valley Oak (*Q. lobata*), Blue Oak (*Q. douglasii*), Engelmann Oak (*Q. engelmannii*), and live oaks such as Coast Live Oak (*Q. agrifolia*) and Interior Live Oak (*Q. wislizenii*). Other trees present may be Digger Pine (*Pinus sabiniana*), with its peculiarly forked trunks and massive cones, the attractive California Buckeye (*Aesculus californica,* Hippocastanaceae), and the Southern California Walnut (*Juglans californica,* Juglandaceae). Understory plants are species that occur also in Valley Grassland or in Chaparral.

Valley Grassland (Plates 9D; 10A, B; Map 2)

In many respects the Valley and Foothill Woodland may be considered as a plant community that is transitional between the true forest communities (such as North Coast Forest or Montane Forest) of upland areas or coastal moist areas and the treeless grassland communities of the valleys represented in California by the Valley Grassland. The Valley Grassland plant community occupies (or occupied) most of the floor of the Central Valley. This plant community is one which has been greatly reduced in size in the past two centuries, since it occupied lands that have been transformed into agricultural land. As a result, Valley Grassland occupies only a small remnant of its former area.

Originally, Valley Grassland was made up of various perennial bunch grasses (Plate 9D) such as needle grass (*Stipa* spp.), bunch or blue grass (*Poa* spp.), and three-awn (*Aristida* spp.). These grasses have completely disappeared in large areas of the Central Valley where the native grass cover has been removed and the land has been planted with cultivated crops, or where destructive sheep or cattle are pastured. Much of the Central Valley is still grassland, but even in grazed areas the cattle or sheep have exterminated the native perennial grasses, and these have been replaced by introduced annual grasses such as brome grass (*Bromus* spp.), wild oats (*Avena* spp.), and fescue (*Festuca* spp.). The golden hills that characterize much of California are golden in the summertime because of the dry stems and leaves of these introduced annual grasses; it is probable that when these areas were occupied by the native perennial grasses they were green or gray-green during most of the year.

Valley Grassland occurs most extensively in the Central Valley, but also is present in some of the low valleys or gentle slopes of the Coast Ranges and in some areas of the Transverse and Peninsular Ranges. It also occurs along the coast from San Luis Obispo County southward to the Mexican border. Rainfall in the Valley Grassland is variable, but is generally less than 20 inches (51 cm) per year. This low rainfall probably is responsible for the absence of trees in the Valley Grassland. Summer temperatures may be very high, and heavy frosts are common in some areas in the winter.

In the spring, portions of the Valley Grassland are covered by a rich array of spectacularly colorful spring annuals (Plate 10A). Areas that are well known to professional and amateur botanists include the low hills and valley bottoms in the Bakersfield area and the Tehachapi foothills, the Solano delta area, the Red Bluff region, and some of the interior valleys of the Coast Ranges. In years in which there is abundant rainfall, hundreds or thousands of acres of Valley Grassland are occupied by masses of these showy annuals.

Of particular interest in the Valley Grassland is the type of habitat known as a vernal pool, sometimes locally called hog wallows. These pools occupy depressions in the grassland area that fill with water during the winter. As the pools begin to dry up in the spring, various annual plant species begin to flower (Plate 10B). The result is local patches of color that may persist for relatively long periods of time until the pools dry up. A number of plants are restricted to these vernal pools of the Valley Grassland; these include several species of meadow foam (*Limnanthes* spp., Limnanthaceae), downingia (*Downingia* spp., Lobeliaceae), goldfields (*Lasthenia* spp., Compositae), and other colorful genera.

In Miocene and Pliocene times the area now occupied by the Central Valley was a large inland sea. This sea diminished subsequently, although it persisted into recent times in the form of extensive lakes and marshlands that occupied the Central Valley well into this century. One consequence of this relatively recent availability of the valley floor for occupation by plants is that many of the plant species that are restricted to the Central Valley are of recent evolutionary origin: the evolution of these species was associated with the appearance of a new and ecologically distinctive land area for occupation by land plants. The surrounding upland areas have been occupied by plants for a much longer period of time than the Central Valley and support evolutionarily older plant species and plant communities.

Riparian Woodland (Plate 10C, D)

Since the climatic regime over much of California is an arid one, the local occurrence of permanent standing or running water has a striking influence on the vegetation. The many large streams and rivers that flow out of the California mountains are generally lined with deciduous trees, shrubs, and herbs that are restricted to the banks of these water courses. Often these plants follow the streams or rivers out into the Valley Grassland. In various portions of California stream- and riverbanks are occupied by trees such as Bigleaf Maple (*Acer macrophyllum*), Black Cottonwood (*Populus trichocarpa*), and White Alder (*Alnus rhombifolia*). At lower elevations and on the valley floors, water courses are lined with

California Sycamore (*Platanus racemosa*), California Boxelder (*Acer negundo* subsp. *californicum*), and Fremont Cottonwood (*Populus fremontii*), along with a number of species of willows (*Salix* spp.). Where river valleys are broad, the extent of the Riparian Woodland is often correspondingly broad; in other areas, particularly at higher elevations where the water courses are narrow and the streambanks are relatively steep, Riparian Woodland may form a very narrow strip of forest that is only a few meters in width. The distribution of this plant community has not been included on the map of Woodlands, but it can be plotted easily on a map that shows the occurrence of year-round rivers and streams in the state.

Freshwater Marsh (Plate 11A; Map 2)

In areas of the state where there are fairly large expanses of standing or very sluggishly moving, shallow water, one generally encounters the Freshwater Marsh plant community. Floristically, this plant community is a relatively simple one whose main components are various species of cattails (*Typha* spp., Typhaceae), bulrush or tule (*Scirpus* spp., Cyperaceae), and sedges (*Carex* spp., Cyperaceae). Freshwater Marsh occurs in the Central Valley along river courses, creeks, and sloughs, or in the vicinity of lakes such as Tulare Lake; extensive marshes also occur in the Sacramento-San Joaquin delta area. Marshlands are present in some inland areas such as Sierra Valley north of Lake Tahoe and in the Modoc Plateau area east of the Sierra-Cascade axis.

The characteristic plant species of the Freshwater Marsh plant community are mostly monocots with a superficial grasslike appearance. Nearly all of the species are perennials that have excellent means of vegetative propagation. One immigrant cattail in a marshy area can occupy many square meters of marsh in a rather short period of time due to rapid vegetative increase, and in many areas of the state open waterways may become clogged by the masses of plants resulting from this growth.

Accounts of early 19th century explorers in the Central Valley indicate the great difficulty that these people had in getting their horses across the valley because of the marshlands. Thomas Coulter, an Irish physician and botanist who visited California in

100

the 1830's, mentioned that the accounts of the large size of the lakes in the Central Valley were "exaggerated"; neither of the lakes, he stated, was over 100 miles (161 km) long! However, the Freshwater Marsh plant community has become considerably more restricted in its distribution than it was in former times. Much of the decrease in marshlands in the Central Valley has been in response to the increasing aridity of the California climate, but the natural process has been overshadowed by the draining of marshes by man in order to increase the expanse of land available for agriculture, by withdrawing ground water for irrigation purposes, and by the diversion or retention of water by dams in the Sierra and the Coast Ranges.

Montane Forest (Plate 11B-D; Map 4)

In continuing the eastward transect across northern California, as we move from the grasslands and marshes of the Central Valley upward into the foothills of the Sierra Nevada, the first plant community encountered is the Valley and Foothill Woodland, which was discussed earlier since it also occurs in the Coast Ranges as well. After leaving the Valley and Foothill Woodland, with a further increase in elevation coniferous trees begin to appear in abundance. At this point we have entered the Montane Forest. In the Sierra Nevada, the Montane Forest begins at approximately 2,000 feet (610 m) in elevation (depending on latitude and local ecological conditions). It may extend up to 8,000 feet (2,438 m) in the southern California mountains. Montane Forest also occurs in the North Coast Ranges and the Klamath-Siskiyou region, and extends into northeastern California. Perhaps the most common and conspicuous conifer in the Montane Forest is Ponderosa Pine (*Pinus ponderosa*). Also present are Incense Cedar (*Calocedrus decurrens*), White Fir (*Abies concolor*), Douglas Fir (*Pseudotsuga menziesii*), Sugar Pine (*Pinus lambertiana*), and in some areas Coulter Pine (*P. coulteri*). In some portions of the Sierra Nevada, the Montane Forest is locally dominated by Sierra Big Tree (*Sequoiadendron giganteum*, Plate 11D) which, like its coastal cousin, is a narrow endemic with a long fossil record. The present scattered distribution of Big Tree can be traced back to the patterns of Pleistocene glaciation several

101

thousand years ago. This massive tree species has not been successful at reoccupying its former pre-Pleistocene range. In southern California, a component of Montane Forest is Bigcone Spruce
(*Pseudotsuga macrocarpa*), a close relative of Douglas Fir but a
species that is adapted to more arid conditions than Douglas Fir.
A few deciduous hardwoods are associated with the coniferous
trees of the Montane Forest. Also present are Canyon Oak (*Quercus chrysolepis*) and California Black Oak (*Q. kelloggii*). At the
upper, cooler, and wetter margin of the Montane Forest, Jeffrey
Pine (*Pinus jeffreyi*, Plate 11B), Red Fir (*Abies magnifica*, Plate
11C), and Lodgepole Pine (*P. murrayana*) begin to appear.

Because of the large area occupied by the Montane Forest and
the diversity of ecological niches that it spans, one might expect
a large number of shrubby and herbaceous plant species to occur
in this plant community, and according to Munz' *A California
Flora*, approximately 1,200 herbaceous plant species occur in
the Montane Forest, and somewhat over 200 shrub species occur
in the understory of the Montane Forest. It is therefore difficult
to give a list of representative understory species, but a few genera or species of wide distribution in the forest community can
be mentioned. These include Mountain Misery (*Chamaebatia
foliolosa*), a low, rather attractive but malodorous member of
the Rose family that forms large masses under the conifers. Also
present are various species of gooseberry or currant (*Ribes* spp.,
Grossulariaceae), blackberries (*Rubus* spp., Rosaceae), manzanitas (*Arctostaphylos* spp., Ericaceae), and California lilacs (*Ceanothus* spp., Rhamnaceae).

One of the tree species present in the Montane Forest is Sugar
Pine (*Pinus lambertiana*). When California was first settled by
immigrants from the eastern states, Sugar Pines were much more
common in the Sierra Nevada than they now are. This former
abundance was the result of the fact that forest fires raged un-
checked through many areas of the Sierra, and Sugar Pine was
perpetuated as a result of these fires. Seedlings of this species
compete successfully only in areas that have been opened up by
fires (or some other ecological disturbance). With the subsequent
effective fire control that has existed in much of the forested
area of the Sierra, natural succession has resulted in the reduction

102

in numbers of Sugar Pines, which have been replaced in the ecological succession by Incense Cedar and White Fir. It is probable that several of the other tree species in the Montane Forest are also "fire type" trees, and their numbers may also be changing as a result of the efficiency of the fire control activities of federal, state, and private agencies.

Montane Chaparral (Plate 12A, B)

At moderate to high elevations in the mountains, particularly the Sierra Nevada, the coniferous forest may be interrupted by areas of Montane Chaparral. This plant community resembles in general aspect the Chaparral of lower elevations, but is given separate recognition because the number and identity of the species of shrubs that occur in the two plant communities are different. Although manzanitas (*Arctostaphylos* spp.) and California lilac (*Ceanothus* spp.) occur in both plant communities, different species are represented in each of them. The regions in which Montane Chaparral occurs receive considerably higher precipitation than those occupied by Chaparral, but with few exceptions the shrubs maintain a strong xerophytic appearance. In many areas of the mountains, Montane Chaparral is successional in nature, and develops on previously forested sites after forest fires have eliminated the trees (Plate 12B). Because of its sporadic distribution, this plant community was not included on the maps.

Subalpine Forest (Plate 12C, D; Map 4)

Immediately above the Montane Forest, and not sharply differentiated from it, is the Subalpine Forest plant community. Some of the coniferous tree species characteristic of the upper reaches of the Montane Forest also occur in the lower reaches of (or throughout) the Subalpine Forest, with the result that there is a gradual transition between the two forest communities. The Subalpine Forest occurs above the Montane Forest in the Sierra Nevada and is present to a lesser extent in the Cascades in the Lassen-Shasta vicinity. The elevation at which it occurs is variable; at its lowest it occurs between 5,000 and 6,000 feet (1,524 to 1,829 m). In southern California or the desert ranges (such as the

103

the White Mountains) it may extend up to 11,000 feet (3,353 m). The climatic regime in the Subalpine Forest is variable, although in general it is somewhat more rigorous than that of the Montane Forest. Winters are usually very cold and associated with heavy precipitation; the winter snows may provide the explanation for the conical form of subalpine conifers, which inspired the idea for the original A-frame building construction.

The Subalpine Forest is as rich in its number of coniferous dominants and its local diversity as is the Montane Forest. One dominant conifer is Lodgepole Pine (*Pinus murrayana*), which has a sporadic distribution. This relatively small conifer seems to occur in areas in which the local climatic or soil conditions are rather unfavorable for the full development of other coniferous species. Other Subalpine inhabitants are Western White Pine (*Pinus monticola*) and Mountain Hemlock (*Tsuga mertensiana*, Plate 12C). At timberline, one finds gnarled and windswept individuals of Whitebark Pine (*P. albicaulis*, Plate 12D). Other pine species are the closely related trio consisting of Limber Pine (*P. flexilis*), Foxtail Pine (*P. balfouriana*), and Bristlecone Pine (*P. aristata*). Bristlecone Pine occurs (in California) in the very dry White Mountains, where it receives an average precipitation of about 12 inches (30 cm) per year; the trees are widely spaced and not especially tall. Despite its unfavorable environment, Bristlecone Pines have a remarkably long life span and are perhaps the longest-lived organisms on earth. One individual shows growth rings which suggest that it is 4,600 years old, an age exceeding that of the oldest Big Trees. Even the leaves of the Bristlecone Pine are long lived; needle life has been estimated to be as much as 30 years.

In many areas occupied by the Subalpine Forest, particularly at higher elevations, trees are widely scattered. The spaces among the trees are frequently occupied by a number of colorful herbaceous perennials and shrubs such as various wild currants (*Ribes* spp.), willows (*Salix* spp.), and huckleberries (*Vaccinium* spp.).

Montane Meadow (Plate 13A)

Where there is shallow or continuously moist soil in upper montane areas, treeless meadows are encountered that are occupied by a distinctive array of perennial herbs. These meadows support the Montane Meadow plant community, which is characterized by a variety of perennial grasses, sedges, and a number of low, broad-leaved herbs. Also present is the tall, striking Corn Lily (*Veratrum californicum*). In mid- to late summer, the wildflower displays in these meadows are often very colorful. Because of the irregular distribution of this plant community, it was not included in the maps.

Alpine Fell-field (Plate 13B-D)

Despite the ability of pines and other conifers to survive and even prosper under the unfavorable environmental conditions at high elevations, there is a point of elevation at which the climatic regime is too stringent for successful growth of trees of any kind. The point at which this ecological circumstance develops is called timberline and is marked by a sharp reduction in number and size of trees. Above this zone of dwarfed and stunted trees is another plant community which occupies very high elevations in the mountains. This is the Alpine Fell-field plant community, a community of low perennial plants that occurs above 9,500 feet (2,896 m) in the Sierra Nevada (primarily), and also in parts of the Cascades and the San Bernardino and San Jacinto mountains. Precipitation in the Alpine Fell-field occurs mostly as snow in the winter, since the elevation is above the point of maximum rainfall on the slopes of the mountains. In many respects, this plant community occupies an alpine desert. The growing season is less than two months long — sometimes as short as six weeks — and heavy frosts may occur during almost any night of the summer. The appearance of this plant community is much like that of a rock garden. The rock-strewn peaks, slopes, or open fields are frequently covered by masses of perennial herbs that form a low, dense turf. Many of the plants in these areas form small cushionlike mats with densely packed leaves. A few grass and sedge species occur in the Alpine Fell-field, but these are accompanied by a large variety of dicots belonging to diverse genera in several families. Many of the dwarfed inhabitants of the Alpine Fell-field (such as the crucifers, composites, penstemons, phloxes (Plate 13C), wild buckwheats (Plate 13D), and

105

eximium — the Sky Pilot) have large, brilliantly colored flowers and produce a mosaic of brilliant colors in late summer. The distribution of this plant community was not mapped.

Pinyon-Juniper Woodland (Plate 14A; Map 4)

The upper reaches of California mountains are occupied by a treeless plant community, and certain peaks are sufficiently high that they support no vascular plants at all. However, as we pass over the crest of the Sierra Nevada on our transect and drop down along the eastern slopes of these mountains, trees again begin to be evident. The species of trees and the localities in which they appear vary, but on the average we can expect to find Pinyon-Juniper Woodland as the first "forest" community encountered on the eastern slopes of the Sierra Nevada as we descend in elevation from the crest. Pinyon-Juniper Woodland represents a transition between forest and non-forest plant communities in some areas, although in a less consistent way than does the Valley and Foothill Woodland in cismontane California. In some areas, Pinyon-Juniper Woodland is on the drier side of Montane or Subalpine Forest, but in some localities — such as in the desert mountains — Pinyon-Juniper Woodland is bordered above and below by scrub communities of various kinds.

Pinyon-Juniper Woodland takes its name from the Single-leaf Pinyon Pine (*Pinus monophylla*) and various junipers such as California Juniper (*Juniperus californica*), Utah Juniper (*J. osteosperma*), and Western or Sierra Juniper (*J. occidentalis*). These trees are not very tall and occur as fairly widely scattered individuals in this plant community. Also present are Desert Scrub Oak (*Quercus turbinella*), Mountain Mahogany (*Cercocarpus ledifolius*, Rosaceae), Antelope Brush or Bitter Brush (*Purshia tridentata*), and all species of the Sagebrush Scrub plant community, to be mentioned next. Pinyon-Juniper Woodland occurs in the Great Basin and in California mountain ranges from Modoc County into southern California, where it is associated with the Transverse and Peninsular Ranges and some of the desert ranges.

106

Sagebrush Scrub (Plate 14B; Map 3)

A plant community that is often adjacent to Pinyon-Juniper Woodland is the Sagebrush Scrub plant community. This community occurs in relatively deep soils along the eastern base of the Sierra Nevada-Cascade axis from Modoc County southward to the San Bernardino mountains. A few small patches of it occur locally elsewhere. The average annual precipitation ranges from 8 to 15 inches (20 to 38 cm), and much of this falls as winter snow. Summers may be very hot and winters relatively cold. The elevations occupied by this plant community range from about 4,000 feet to 7,000 feet (1,219 to 2,134 m).

Sagebrush Scrub is characterized by the presence of various low, silvery-gray shrubs that are 2 to 6 feet (0.6 to 1.8 m) tall or more. The plant community is named after Basin Sagebrush (*Artemisia tridentata*), a member of the sunflower family (Compositae) which occupies vast areas of the Great Basin. Other species of sagebrush present are the related *A. nova, A. arbuscula,* and *A. cana.* Associated shrubs are rabbit brush (*Chrysothamnus* spp., Compositae), Blackbush (*Coleogyne ramosissima*, Rosaceae), cotton thorn (*Tetradymia* spp., Compositae) and a few other superficially similar shrub species.

6. PLANT COMMUNITIES OF SOUTHERN CALIFORNIA

Foregoing discussions were concerned with the major plant communities that were encountered along a transect in northern California. Because of the relatively symmetrical and orderly arrangement of the chief mountain ranges in northern California, a transect approach was used: most of the plant communities in this part of the state tend to be distributed in a pattern that is related to climatic patterns, and these in turn are strongly influenced by the position of mountain ranges in a north-south series. We now turn to plant communities restricted to southern California, especially the desert portions of the state. In this region, the topography forms more of a mosaic pattern.

Coastal Sage Scrub (Plate 14C; Map 3)

In some respects, a southern counterpart of the Northern Coastal Scrub is the Coastal Sage Scrub, also called Soft Chaparral. The term counterpart is used because the Coastal Sage Scrub occupies a narrow strip along the coast stretching along the coastward side of the South Coast Ranges (and some of the Peninsular Ranges) into Baja California, in much the same relative position occupied by Northern Coastal Scrub in the northern portion of the state. But although the general aspect of the two communities is similar, there is little floristic similarity between the Northern Coastal Scrub and the Coastal Sage Scrub. The Coastal Sage Scrub occurs on rather dry, often steep, gravelly or rocky slopes below 3,000 feet (915 m). Climatically, the area occupied by this plant community is rather mild and has an average of 20 inches (51 cm) of rainfall per year or less. The "scrub" refers to the fact that the major plant species found in the community are shrubby species one to six feet (1.3 to 1.8 m) tall, although a few of the component species are considerably larger than this and might be considered small trees.

The name of this plant community comes from the presence of *Salvia* species such as Black Sage (*S. mellifera*) and Purple or White-leaved Sage (*S. leucophylla*, Labiatae). Other shrubs present are the Coastal Sagebrush (*Artemisia californica*, Compositae); Wild Buckwheat (*Eriogonum fasciculatum*, Polygonaceae); and Coyote Brush (*Baccharis pilularis* var. *consanguinea* — also found in Northern Coastal Scrub). Larger species are the handsome Lemonadeberry (*Rhus integrifolia*, Anacardiaceae) and its toxic relative, Poison Oak (*R. diversiloba*).

Shadscale Scrub (Plate 14D; Map 3)

Most herbaceous plant communities of California are best developed in cismontane northern California. Likewise, the Woodland (or Forest) communities also are more extensive in the northern portion of the state (i.e., north of the Transverse Ranges) than in the south. Examination of patterns of distribution of the scrubland communities, however, indicates that these are better developed in southern California than in the northern part of the state. The Shadscale Scrub plant community is named after one of the dominant species, Shadscale. This is *Atriplex confertifolia* (Chenopodiaceae), an erect, rigidly branched, spiny shrub with rather crowded, round leaves that resemble fish scales. (Curiously, Munz and Keck do not list Shadscale as the common name for this shrub, even though this name is widely used for the shrub in much of the Great Basin and it gave its name to the plant community in which it occurs.) Other members of this desert plant community are Hop Sage (*Grayia spinosa*, Chenopodiaceae), Winter Fat (*Eurotia lanata*, Chenopodiaceae), Spiny Sagebrush (*Artemisia spinescens*, Compositae), matchweed (*Gutierrezia* spp., Compositae), Cheese Bush (*Hymenoclea salsola*, Compositae), Blackbush (*Coelogyne ramosissima*, Rosaceae), and the peculiar gymnospermous shrub, Mormon tea (*Ephedra* spp., Ephedraceae).

Despite the fact that the characteristic shrubs of Shadscale Scrub belong to several plant families that are taxonomically unrelated, there is a strong superficial similarity among them. The shrubs are rather small, seldom over half a meter tall. Generally, they are grayish, small leaved, much branched, and sometimes spiny, and produce smallish flowers. Shadscale Scrub occurs in very heavy, often alkaline (pH 8 to 10) or saline soils that are frequently underlain with a hardpan. The Shadscale Scrub plant community occurs on mesas and flatlands at elevations of 3,000 to 6,000 feet (914 to 1,829 m) in various parts of the transmontane desert regions such as the Owens Valley and the Mojave Desert. It is uncommon in the Colorado Desert areas, which are occupied chiefly by Creosote Bush Scrub. Rainfall in these areas is very low — averaging less than 7 inches (17.8 cm) per year. Summers are dry and hot.

Perhaps the most interesting feature of this plant community is the peculiarity of the common names given to its component shrubs. Otherwise, as Munz and Keck state, it covers "large monotonous areas between Creosote Bush Scrub and Joshua Tree Woodland", both of which communities are clearly esthetically preferable to the Shadscale Scrub.

Alkali Sink Scrub (Plate 15A-C; Map 3)

Another scrub community of arid regions is the Alkali Sink Scrub plant community. This community is made up of halophytic shrubs that belong largely to the Goosefoot family (Chenopodiaceae), a family whose members are salt-tolerant. Characteristic species of the Alkali Sink Scrub plant community are saltbush (*Atriplex* spp.), Iodine Bush (*Allenrolfea occidentalis*, Plate 15C), pickleweed (*Salicornia* spp.), Greasewood (*Sarcobatus vermiculatus*), and seep weed (*Suaeda* spp.). All these genera are members of the Chenopodiaceae. Note that *Suaeda* and *Salicornia* are also present in the Coastal Salt Marsh plant community, which is not surprising in view of their high salt tolerance.

The Alkali Sink Scrub occupies the low-lying, poorly drained alkali flats and playas in the San Joaquin Valley (especially around Tulare Lake) and also occurs in similar habitats in transmontane deserts such as those in the Panamint and Death valleys. The rainfall in such areas may be as low as 2 inches (5 cm) per year or less, but the soil is often saturated with highly saline water for much of the year due to seepage into these low areas. Summer temperatures in these desert regions may be excessively high (as they are, for example, in Death Valley). As might be expected in view of their halophytic nature, Alkali Sink Scrub shrubs characteristically have fleshy leaves and stems. Outside California, this plant community is especially well developed in the low areas around the Great Salt Lake in Utah.

Because these low desert areas are frequently inundated with water during the rainy season, a hard saline crust may form on the soil surface when the soil is dry. In association with this soil phenomenon, a curious symbiotic relationship has developed between local ant species and Alkali Sink plants. Ants bury seeds below the surface in the process of carrying them to their nest (or perhaps storing them) for use as food. This procedure results in the seeds being planted below the hard surface crust at a level where the salinity is lower and the moisture conditions are more favorable to germination.

Joshua Tree Woodland (Plate 15D; Map 4)

Another desert community that is well known to many Californians is the Joshua Tree Woodland, a plant community which only marginally deserves the name "woodland". Joshua Tree is a handsome tree-like yucca, *Yucca brevifolia* (Agavaceae). Common associates of Joshua Tree are Mojave Yucca (*Y. schidigera*), junipers (*Juniperus* spp.), Mormon tea (*Ephedra* spp.), cotton thorn (*Tetradymia* spp., Compositae), California Buckwheat (*Eriogonum fasciculatum,* Polygonaceae), Bladder Sage (*Salazaria mexicana,* Labiatae), box thorn (*Lycium* spp., Solanaceae), and many species of the cholla cactus (*Opuntia* spp.). Rather few of the woody plant species generally considered to be members of this plant community are restricted to it.

Joshua Tree Woodland occupies well-drained mesas and desert slopes from Owens Valley to the Little San Bernardino mountains and southern Nevada and Utah. It occurs at moderate elevations from somewhat over 2,000 feet (610 m) to about 6,000 feet (1829 m). The average annual rainfall is between 6 and 15 inches (15 and 38 cm) depending on locality. Unlike most of the lowland plant communities in California, Joshua Tree Woodland receives occasional summer showers. The individual Joshua Trees and associated junipers are rather widely spaced and are seldom over 30 feet (9 m) high. Numerous shrubby plants in addition to those listed above occur among the trees, and during the spring following a wet winter the ground among the shrubs and trees is carpeted with spectacular masses of showy annuals in flower.

Creosote Bush Scrub (Plate 16; Map 3)

The last plant community to be discussed is the one which is the most widespread in the southern desert portions of California. This is the Creosote Bush Scrub. Creosote Bush is *Larrea divaricata* (Zygophyllaceae), a rather attractive, tall shrub that dominates much of the desert landscape below 3,500 feet (1,067 m) from Inyo County southward (Plate 16A). It also occurs locally in some interior cismontane valleys such as at Poso Creek, Tulare County, and localities in western Riverside County. Also present as plant associates of the Creosote Bush are Burro Weed (*Franseria dumosa*, Compositae), the colorful and spiny Ocotillo (*Fouquieria splendens*, Fouquieriaceae), Brittle Bush (*Encelia farinosa*, Compositae), Cheese Bush (*Hymenoclea salsola*, Compositae), and prickly pears and chollas of the cactaceous genus *Opuntia* (Plate 16B). Because of the limitations of water supply in the area occupied by the Creosote Bush Scrub, the water courses (which are dry most of the year) support a characteristic flora that takes advantage of the abundant supply of water during rainy periods of the winter or summer (Plate 16D). Some botanists would consider this wash woodland to represent a separate plant community. Certain desert trees and shrubs generally occur only along these water courses. These include Palo Verde (*Cercidium floridum*, Leguminosae), Smoke Tree (*Dalea spinosa*, Leguminosae), Catclaw (*Acacia greggii*, Leguminosae), Desert Willow (*Chilopsis linearis*, Bignoniaceae), Chuparosa (*Beloperone californica*, Acanthaceae), and Desert Lavender (*Hyptis emoryi*, Labiatae). Another interesting tree that occurs around moist, somewhat alkaline spots in the Creosote Bush Woodland is California Fan Palm (*Washingtonia filifera*), which often coexists with various willows (*Salix* spp.). This species is rather uncommon in nature, although it is widely planted as an ornamental in subtropical regions.

The seeds of many wash woodland tree species are very hard-coated and will not germinate even if left in water for over a year.

112

It is necessary to scratch the coat of these seeds in order for germination to occur, otherwise they are impervious to water. The grinding action of sand and rocks in the flash floods of the desert performs the scarification function and also provides the seedlings with abundant water which will supply their requirements during the first few weeks of growth. Such floods also serve to disperse the seeds. Like many desert perennials, seedlings of wash woodland trees produce only two or three leaves immediately after germination and then seemingly become dormant. However, these plants are far from dormant during this time, but are devoting their chief energies to developing extensive, deep root systems that will enable them to survive long after the moisture from the flood has dissipated.

Summer temperatures in the Creosote Bush Scrub may be very high, and in many areas winter temperatures do not drop to the freezing point. The average annual rainfall in this plant community is very low, ranging from 2 to 8 inches (5 to 20 cm). In appearance, Creosote Bush Scrub is composed of numerous shrubs or small trees to 10 feet (3 m) high or somewhat higher, that are widely and symmetrically spaced. Some of these species, particularly those that occur along the desert washes, are very colorful when in flower.

The Creosote Bush Scrub is dominated by woody plants, but it is also a plant community in which there is a rich representation of annual plant species (Plate 16C). In addition, herbaceous perennials also are present although they are less abundant than are herbaceous annuals. The general vegetational aspect of the Creosote Bush Scrub during almost any month of the year belies its arid nature. It is usually green and, since it supports a cover of shrubs and small trees, it may give the impression that it receives more rain than it actually does.

7. EVOLUTION OF THE CALIFORNIA FLORA

The natural vegetation of California represents a set of plant communities of varied historical origins. We have already pointed out that some plant communities such as the North Coastal Forest (or at least the segment of this community dominated by Coast Redwood) have had a continuity back into the geological past that extends as far as twenty million years, although these plant communities have not always occupied their present geographical range. Other plant communities, such as those that are adapted to the characteristic Mediterranean or desert climate of California, are of relatively recent origin because the climatic regime under which they exist is one of recent origin.

The following geological time scale will be of some help in visualizing the past history of the rich flora that now occupies the California Floristic Province:

EPOCH	STARTED
Recent	post-glacial
Pleistocene	1 million years ago
Pliocene	10 million years ago
Miocene	25 million years ago
Oligocene	40 million years ago

The Oligocene, Miocene, and Pliocene along with the older Eocene and Paleocene epochs, constitute the Tertiary period. In Oligocene and Miocene times the area occupied by California and much of the rest of the western United States was a region of rolling plains or low mountains. There was no major mountain system in California until the close of the Tertiary period.

The Arcto-Tertiary Geoflora

As a consequence of topographical uniformity in the early Tertiary, the climate of the western part of North America was much less varied than it is at the present time. It is not surprising, therefore, that much of North America, northern Asia, and Europe was covered by a rather uniform type of vegetation, and a rich one in terms of number of species present. In the west,

114

this flora extended from about the latitude of San Francisco well northward into what are now arctic regions. This forest has been termed the Arcto-Tertiary geoflora. "Arcto" refers to its northern distribution; "Tertiary" refers to the time period during which it flourished; and a "geoflora" is a major vegetation unit that has continuity in space and time. In recent years, the concept of the Arcto-Tertiary geoflora has been a subject of some dispute; nevertheless, it does help to bring home the idea that the contemporary flora of California has had a long history in time and space.

The Arcto-Tertiary geoflora contained a number of tree genera that have persisted on the Pacific coast until the modern day. These include such genera as *Tsuga* (hemlock), *Thuja* (Canoe or Western Red Cedar), *Picea* (spruce), and deciduous trees such as maples (*Acer*) and dogwoods (*Cornus*). However, in response to climatic changes that occurred over a long period of time, other genera in this forest disappeared from the western part of the continent, although they still persist in the eastern part of the continent. These include such genera as beech (*Fagus*), chestnut (*Castanea*), elm (*Ulmus*), sweet gum (*Liquidambar*), and some other hardwood genera that make up the modern eastern deciduous forest. It is interesting that the remnants of the Arcto-Tertiary geoflora that persisted in the west are largely coniferous genera — the deciduous genera mostly disappeared from this region — while in the eastern portion of the continent the reverse was true — the conifers largely disappeared and the hardwoods remained. Some genera that still occur in either the eastern or the western forests of North America also survived in the Old World, although there they are now represented by different species from those present in North America. Other genera, such as *Ginkgo* (the maidenhair tree) and *Ailanthus* (the tree of heaven) disappeared completely from North America, whereas others, such as *Sequoia*, disappeared completely from the Old World.

Since the middle of the Miocene there have been three major climatic and geological changes in the western United States that have been responsible for the elimination of the old Arcto-Tertiary forest over much of the area and its replacement by other plant communities:

115

1. Since Tertiary times winter rainfall has become reduced and summer rainfall has essentially disappeared over much of California.

2. Starting in the Pliocene, the Sierra Nevada and the Cascade Range were uplifted. The result of this major geological event was the formation of a rain shadow to the east of these mountains and the appearance of new upland areas for plant colonization, particularly on the western slopes. Furthermore, the western slopes of these mountains are now comparatively well watered with rain and snow.

3. During Pleistocene times, much of northern North America was covered with glacial ice sheets. In the west, these barely extended southward across what is now the U.S.-Canadian border, but because of the cooler climatic conditions that prevailed on the continent during this time, many of the upper reaches of the Sierra Nevada in California were covered with glaciers of various sizes. In addition, there was a general cooling trend over much of western North America.

The biological consequences of these geological and climatic features are complex. The development of a cooler climate and the action of the glaciers in mountainous areas clearly had their effect in eliminating certain species from California and in altering the distribution ranges of other species. As an example, Sierra Big Tree (*Sequoiadendron giganteum*) now occurs only in areas of the western slopes of the Sierra Nevada that were not devastated by the action of glaciers. The reduction in total annual rainfall to the east of the Sierra Nevada eliminated the forest climate and replaced it with a scrubland or grassland climate. The rich Arcto-Tertiary forest was rapidly eliminated from much of the Great Basin because of this climatic change. However, the relatively conservative and moist climate that remained along certain areas of the western slopes of the Sierra Nevada provided pockets in which some of the Arcto-Tertiary forest species took refuge. It was from these pockets that a recolonization of the Sierra Nevada took place following the retreat of the last Pleistocene glaciers and the subsequent warming of the climate. It was

116

also in relatively recent times that the zonation of forest communities took place on the western slopes of these mountains. It has been suggested that the major reorganization of the Arcto-Tertiary forest in California occurred on the western slopes of the Sierra Nevada and that the forest communities that developed in this area subsequently invaded the Coast Ranges, the mountains of southern California, and the mountains of Baja California. It has even been suggested that some species of conifers now considered typical of the North Coast Ranges (such as Coast Redwood, Port Orford Cedar, and Canoe Cedar) are probably fairly recent immigrants into this area from the western slopes of the Sierra Nevada, where they have subsequently become extinct.

Many of the modern plant communities in California contain vestiges of the Arcto-Tertiary geoflora. It seems probable that some of these communities represent local aggregations of species whose immediate ancestors occurred with the ancestors of species that are present in other remnant plant communities. In recent times there has been a segregation of the descendants of Arcto-Tertiary tree species into smaller, more homogeneous plant communities than were occupied by their forebears. Some of these derivative plant communities (such as the Closed-Cone Pine Forest) are not very successful in coping with contemporary biotic and climatic conditions of the Pacific coast and seemingly are on their way to complete disappearance as communities. Other of these communities (such as most phases of the North Coastal Forest) seem to be well adapted to current ecological conditions and are therefore successful.

The Neotropical Tertiary Geoflora

At the southern edge of its range, the Arcto-Tertiary geoflora merged into the Neotropical Tertiary geoflora. This second geoflora covered the southern portion of North America, although its boundaries with the Arcto-Tertiary geoflora oscillated in response to long-term climatic changes. In the Eocene, for example, segments of the Neotropical Tertiary geoflora extended as far north as Alaska, at which time the Arcto-Tertiary geoflora must have been squeezed into a rather small area along the northern

117

fringe of North America. The Neotropical Tertiary geoflora was composed of diverse tropical or subtropical trees such as figs, avocados, cinnamon, palms, and others. This geoflora has now disappeared from most of North America, although it is represented in the region of southern Mexico southward to northern South America. This geoflora is now restricted to tropical areas with a high rainfall. Its disappearance from much of its former area was due to the cooler, drier climates that have prevailed over much of North America since Eocene times. There are few living remnants of this rich flora left in California, although some genera managed to survive here into the Pliocene.

The description given above of the former and present ranges of the Arcto-Tertiary and Neotropical Tertiary geofloras and their response to climatic changes suggests that there is a large area of the southwestern portion of North America that is generally unfavorable for occupancy by the modern descendants of either one of these geofloras. Yet this southwestern area is now occupied by a rich flora. What is the origin of these plants?

The Madro-Tertiary Geoflora

According to D. I. Axelrod of the Department of Botany, University of California, Davis, in middle Miocene times the countryside south of San Francisco was occupied largely by an oak woodland flora that had migrated into this area from the Sierra Madre Occidental region in northwestern Mexico. The plants of this oak woodland were adapted to a year-round rainfall but to a warmer, somewhat drier climate than were the plants of the Arcto-Tertiary forest to the north of it. Presumably this oak woodland occupied some of the area that formerly had been occupied by the Neotropical Tertiary geoflora, which in turn had become gradually eliminated from the American southwest due to climatic changes. It is probable that some species of this oak woodland evolved directly from subtropical precursors in the Neotropical Tertiary geoflora. This oak woodland has been termed by Axelrod the Madro-Tertiary geoflora, the "Madro-" coming from the name of the Mexican mountains (Sierra Madre Occidental) which now occupy the general area from which this flora is believed to have migrated northwestward.

The Madro-Tertiary geoflora consisted largely of sclerophyllous, small-leaved trees and shrubs similar to those that presently occupy many areas of California. Even as early as the middle Eocene some of the immediate ancestors of species present in Chaparral, Coastal Sage Scrub, and Valley and Foothill Woodland were present in California, although it is doubtful if these plant communities would have been recognizable at that time. During the middle Pliocene times, the oak woodland dwindled and disappeared over many areas of what later became the southwestern deserts, since the rainfall in these areas became sharply reduced as the Sierra Nevada and the Peninsular ranges were uplifted. The increasing dryness that characterizes the climatological history of the Pacific coast is one that has been developing gradually, although erratically, over perhaps as many as a hundred million years. Therefore, during this time there have been plants that have been gradually adapting to these altered moisture conditions; those that did not adapt either became extinct or else persist only in areas where the rainfall is sufficiently high to support them. For example, Catalina Ironwood (*Lyonothamnus floribundus*, Rosaceae) was once present on the mainland of California but presumably became extinct there because of its inability to tolerate the increasingly dry climatic conditions. As a consequence, it now persists as a relictual endemic of the Channel Islands. Likewise, some of the pine species in the Closed-Cone Pine Forest once were also more widely distributed, and these are barely persisting in isolated pockets along the cool and relatively moist Pacific coast. There is evidence that the increasing aridity of the California climate has altered the composition of some plant communities that we consider to be characteristic of the contemporary climate. For example, Chaparral was probably once more widespread than it now is and also contained a larger number of species of shrubs than it now does. There were also areas where the Chaparral contained some trees and also some of the thorn-scrub plants that are now found only to the south of the present range of Chaparral. Such genera as *Acacia* and *Fouquieria* (Ocotillo) once were components of Chaparral, but now

119

these have become sorted out from the Chaparral and occur in another plant community to the south.

Uplift of the mountain ranges had an important effect in creating new habitats because of the cooler climates that prevail in upland areas and because of the increased rain- or snowfall on the western slopes of the mountains. The Montane Forest and the Subalpine Forest plant communities both occupy habitats that were not present before the uplift of the mountains in the Pliocene. However, most of the genera of trees present in both of these plant communities were also present in the Miocene Arcto-Tertiary geoflora and are likely derived from these species. The plant species present in the Alpine Fell-field plant community were not derived from precursors present in the forest communities. Some of these species undoubtedly migrated into their present habitats from similar habitats far to the north. As a consequence, some species in these fell-fields are related to those presently widespread in arctic regions. A second source of derivation of these high altitude plants is from the adjacent deserts; some of the genera of the fell-fields (such as *Eriogonum*) fall into this category. It is not surprising that this desert derivation has occurred, since the fell-field climate is virtually that of a desert, albeit one at a very high altitude.

The three geofloras that have been discussed were complexes of several distinctive plant communities, and their composition and ranges fluctuated considerably over long periods of geological time. On the average, however, the two geofloras that have made major contributions to the contemporary flora of California were very different ones, adapted to different average climatic regimes. As a consequence, the plant communities derived from these two geofloras have sorted themselves out into a fairly well-marked pattern: the Arcto-Tertiary derivatives currently inhabit the mountainous areas and the northern part of the state, both of which are comparatively cool and well-watered; the Madro-Tertiary derivatives occupy the cismontane and the southern portion of the state including the deserts. Although there is seemingly little direct heritage of the Neotropical Tertiary geoflora in contemporary California, it is likely that some of the California representatives of essentially tropical families may

120

be derived from this third geo-flora. These would include such genera as California Bay (*Umbellularia californica*, Lauraceae), flannel bush (*Fremontodendron* spp., Sterculiaceae), and California Fan Palm (*Washingtonia filifera*).

Recent Changes in the Flora

The present-day California vegetation and flora trace their history back several tens of millions of years into early Tertiary times. Man has been in California a brief period of time. However, man himself has been responsible for several drastic and rapid alterations in the flora of California which are paralleled in magnitude only by the long-term changes that occurred slowly in response to geological and climatic changes in past geological times. The changes that the Indians brought about in the native flora are unknown, although the burning practices carried on by several Indian groups undoubtedly had an important local effect in eradicating certain native species and encouraging others. The present distribution of the native northern California walnut (*Juglans hindsii*) is undoubtedly partly due to the California Indians' practice of carrying the edible nuts of this species from place to place. Many individual specimens or small colonies of this rather uncommon tree are located at the sites of former Indian campsites in the region to the north and east of San Francisco Bay, although the degree to which the present distribution of this tree has been influenced by the activities of Indians is still a matter of contention.

Another change in the flora has been the extinction by man of a few plant species and the reduction of other species to a few or perhaps only a single known population as a result of urbanization or development of agricultural lands. Many rare species, such as Large-Flowered Fiddleneck (*Amsinckia grandiflora*, Boraginaceae), Hickman's Mallow (*Sidalcea hickmanii*, Malvaceae), Catalina Ironwood (*Lyonothamnus floribundus*, Rosaceae), and Pale Brodiaea (*Brodiaea pallida*, Amaryllidaceae), have been rare since they were first brought to the attention of biologists and are rare for natural rather than man-made causes. Such restricted species naturally attract the attention of conservationsts. Yet a

121

century ago, who would have predicted that there was danger to the thousands of acres of Valley Grassland? Many species in this area have been radically reduced in numbers, and although perhaps few of them face immediate extinction, they certainly should be placed on the list of endangered plant species.

The eradication of many natural plant communities by man has been accompanied by the replacement of these communities with cities, freeways, suburbs, or other sorts of developed land which are essentially devoid of vegetation. However, another more subtle alteration of the California flora has resulted from the transformation of land occupied by native plants into grazing land. Approximately thirty million acres of land in California are cultivated or grazed. This land is largely occupied by agricultural plants that require cultivation in order to survive. Grazing lands and other lands which have been less severely disturbed often are occupied by numerous species of introduced plants that have become established as a part of the alien flora of the state. About one-eighth of the present flora of the state consists of plant species that have been introduced into the state from elsewhere in the last 200 years. These are plants that mostly are classified as "weeds". A gardener or agriculturalist might define a weed as a plant that is growing where it is not wanted; a botanist might add more specific biological qualifications to his definition. Weeds possess several characteristics that enable them to survive in disturbed habitats and, indeed, a number of weed species can survive only in disturbed areas. Weeds, like cultivated crops, require man-made conditions in order to survive, although weeds are not encouraged but survive as unwanted camp followers, often thriving under the same conditions as cultivated plants.

How did these weedy plants arrive? Some of them were introduced into California consciously because they had ornamental value. For example, the European Sweet Alyssum (*Lobularia maritima*, Cruciferae) was introduced as a garden plant because of its drought tolerance and attractive, scented white flowers. It has become established locally as an inhabitant of poorly kept gardens and empty lots. Likewise, Red Valerian (*Centranthus ruber*, Valerianaceae) was introduced into the state as an ornamental, but has become locally established in waste places

122

and along roadsides. The leaf stalks of Cardoon (*Cynara cardunculus*, Compositae) are prized as a food item by people of Mediterranean ancestry; this attractive thistle has escaped from gardens and has become locally established on the hills around San Francisco Bay. This is perhaps also the explanation for the presence of Sweet Fennel or Finocchio (*Foeniculum vulgare*, Umbelliferae), which is common in waste places of central and southern California. The broad definition of a weed also can include trees such as Blue Gum (*Eucalyptus globulus*, Myrtaceae), a tree which is maintaining itself in some areas of California where it was introduced as an ornamental and for possible lumber purposes. Several species of tamarisk (*Tamarix* spp., Tamaricaceae) have become established along water courses, although the trees were originally introduced as ornamentals or as windbreaks. Perhaps one of the most pernicious weeds in California is Johnson Grass (*Sorghum halepense*, Gramineae), originally grown as a forage grass. It has since become established widely as a troublesome and economically expensive pest of field crops. The list of such weeds that have escaped from cultivation is an extensive one.

It is probable that the majority of weeds in California have resulted from accidental introductions of seeds rather than from intentional introductions. In former days, much of the seed of various field or garden crops planted in California originated from Old World areas and was contaminated with a number of weed seeds from these regions. Planting of the seeds of these crop or ornamental plants resulted in accidental planting of weeds as well. In addition, in former years a number of weeds have been introduced into the American west as contaminants of wool imported from the British Isles, South America, or Australia. Such weeds initially became established in the vicinity of woollen mills, but many of them spread rapidly from these sites of introduction. Another means of weed introduction in the nineteenth and early twentieth century was via ballast. In former times ships visiting the west coast of North America carried ballast in the form of soil or rocks that had been put on board in various foreign ports. Ballast was emptied from the ships at dockside, and a number of unusual and interesting botanical aliens have arrived in California via these ballast dumps.

The first weeds known to arrive in California have been "fossilized" in adobe bricks. Filaree (*Erodium cicutarium*, Geraniaceae), Curly Dock (*Rumex crispus*, Polygonaceae), and Sow Thistle (*Sonchus asper*, Compositae) are all Old World weeds whose seeds have been found in adobe bricks of old Spanish buildings constructed before the Mission Period (1769-1824). Examination of adobe bricks made in the Mission Period indicates that an additional number of weedy species arrived and became established during that time. These include Black Mustard (*Brassica nigra*, Cruciferae), wild oats (*Avena* spp., Gramineae), and Wild Carrot or Queen Anne's Lace (*Daucus carota*, Umbelliferae), all from the Old World. Interestingly, however, the extensive and rich botanical collections of such early plant explorers as David Douglas, Thomas Nuttall, and John Fremont who visited the state in the 1830's and 1840's contained only a single weed species. In view of the fact that some of these men had to carry their own collections and collecting gear on their backs, perhaps they consciously avoided collecting plants that they knew to be introduced, since they were more interested in new and unusual native species that they were continually encountering. In the early 1840's, the Russians made extensive collections in Sonoma County, working out of their settlement at Fort Ross. These collections reveal a varied assortment of introduced weeds, indicating that a number of these had been introduced long before extensive settlement of the area and suggesting that these introduced plants were present but avoided by other collectors.

By 1860 only about 100 species of weeds had been recorded in California, but it is certain that many more had become well entrenched by that time. After 1860, more attention was given to recording the adventive flora of the state, with the result that in the past century the arrival of new weeds and their spread have been fairly well documented. This has been assisted by the watchful eye of various county agents and by biologists in the state Department of Food and Agriculture.

It is difficult to predict whether or not an introduction will be successful and, if so, whether the species will spread from its point of introduction. For example, the yellow-flowered *Uro-*

spermum picroides (Compositae) is a widespread and well known weed in the Mediterranean region. Plants were first recorded in North America from a small colony that had become established near the Engineering Building on the Berkeley campus of the University of California in 1915. A recent visit to this site revealed plants still growing in this area of initial introduction, but this species apparently has not succeeded in spreading from its small foothold in the New World, since it is otherwise unknown in California or elsewhere in the U.S. In contrast, however, some species which are well behaved in their homeland may become widespread and sometimes aggressive weeds in their new home. For example, a relative of the filaree, *Erodium obtusiplicatum* (Geraniaceae), is a rare plant in northwestern Africa, yet is certainly one of the commonest plants in California. Gorse (*Ulex europaeus*, Leguminosae) is a relatively well behaved shrub in its native Europe, but has become a troublesome weed from central California northward into southern Oregon. Gorse burns very easily and rapidly, and as a consequence is a considerable fire hazard in areas where it has become established. In the 1930's much of the town of Bandon in southern Oregon was destroyed by a fire whose destructiveness was due to the large areas of land in and around the town that were (and still are) covered by Gorse.

Most weedy species that occur in California are of exotic origin, although a few native species have also developed weedy tendencies. For example, the fall- and winter-flowering Telegraph Weed (*Heterotheca grandiflora*, Compositae) is a native species that probably originated as an inhabitant of sandy soils in southern California. In recent years this species has spread northward into central and northern California, where it is now common along roadsides or in sandy fields. Another weedy Composite is Pineapple Weed (*Matricaria matricarioides*), an extremely common annual weed of waste places in the west which undoubtedly was originally native to the area, although its natural range is unknown since it is so widely spread as a weed of disturbed habitats. Another genus which has been successful in spawning weeds is the fiddleneck genus *Amsinckia* (Boraginaceae) which is a roadside and grassland weed over much of California.

125

Most of the weedy plant species in California, however, are not derived from native species but have been introduced from elsewhere. About seventy-five percent of the weeds of California are of European and western Asian origin, which is not surprising since these regions have been cultivated and inhabitated for very long periods of time and have offered the proper habitats for the origin of weeds for centuries. About 10 percent of the weeds of California are of South American origin, and the remainder form a miscellany from elsewhere in the world. The Gramineae and Compositae together are responsible for about forty percent of the weed species in the state.

REFERENCES

I. References for various topics covered in the book

Chapter 1

Bell, C. R. 1967. Plant variation and classification. Wadsworth Publishing Co., Belmont. Chapter 1, 2.

Harrington, H. D., and L. W. Durrell. 1957. How to identify plants. Swallow Press, Chicago.

Chapter 2

Billings, W. D. 1970. Plants, man and the ecosystem. Wadsworth Publishing Co., Belmont. Chapter 2, 3.

Gankin, Roman, and J. Major. 1964. *Arctostaphylos myrtifolia*, its biology and relationship to the problem of endemism. Ecology 45: 792-808.

Howell, John T. 1957. The California flora and its province. Leaflets of Western Botany 8: 133-138.

Kruckeberg, A. R. 1954. Plant species in relation to serpentine soils. Ecology 35: 267-274.

Mason, Herbert L. 1946. The edaphic factor in narrow endemism. I. The nature of environmental influences. Madroño 8: 209-226. II. The geographic occurrence of plants of highly restricted patterns of distribution. Madroño 8: 241-257.

McPherson, J. K., and C. H. Muller. 1969. Allelopathic effects of *Adenostoma fasciculatum*, "Chamise", in the California chaparral. Ecological Monographs 39: 177-198.

Noldeke, A. M., and J. T. Howell. 1960. Endemism and the California flora. Leaflets of Western Botany 9: 124-127.

Stebbins, G. L., and J. Major. 1965. Endemism and speciation in the California flora. Ecological Monographs 35: 1-35.

Also see Chapter 1 references.

Chapter 3

Clausen, Jens, D. D. Keck, and W. M. Hiesey. 1940. Experimental studies on the nature of species. I. Effect of varied environments on western North American plants. Carnegie Institution of Washington, Publication 5201.

Dale, R. F. 1959. Climates of the states. California. U. S. Bureau of Commerce Weather Bureau.

Durrenberger, Robert W. 1968. Patterns on the land. (Geographical, historical and political maps of California.) National Press Books, Palo Alto.

Merriam, C. H. 1898. Life zones and crop zones of the United States. U. S. Dept. of Agriculture, Biological Survey Bulletin 10.

Oakeshott, G. 1971. California's changing landscape. McGraw-Hill, New York.

Went, F. W. 1955. The ecology of desert plants. Scientific American, April.

Chapter 4

Critchfield, W. B. 1971. Profiles of California vegetation. U.S. Dept. of Agriculture, Forest Service Research Paper PSW - 76.

Munz, P. A. 1959. A California flora. Univ. Calif. Press, Berkeley and Los Angeles. pp. 10-18.

Munz, P. A., and D. D. Keck. 1949. California plant communities. El Aliso 2: 87-105; 199-202.

Chapter 5

Axelrod, D. I. 1967. Evolution of the California closed cone pine forest. *In* Proceedings of the symposium on the biology of the California islands, R. N. Philbrick, editor. Santa Barbara Botanic Garden.

Cain, S. A. 1944. Foundations of plant geography. Harper & Bros., New York.

Chapman, V. J. 1960. Salt marshes and salt deserts of the world. Interscience Publishers, New York.

Cooper, W. S. 1922. The broad-sclerophyll vegetation of California. Carnegie Institution of Washington, Publication No. 319.

———. 1967. Coastal dunes of California. Memoir 104, Geol. Soc. America, p. 1-131.

Crampton, Beecher. 1974. Grasses in California. Univ. Calif. Press, Berkeley and Los Angeles. Natural History Guide Series.

Dawson, E. Yale. 1966. Seashore plants of southern California. Univ. Calif. Press, Berkeley and Los Angeles. Natural History Guide Series.

———. 1966. Seashore plants of northern California. Univ. Calif. Press, Berkeley and Los Angeles. Natural History Guide Series.

Ferguson, C. W. 1968. Bristlecone pine: science and esthetics. Science 159: 839-846.

Griffin, James R., and William B. Critchfield. 1973. The distribution of forest trees in California. USDA Forest Service, Research Paper PSW - 82.

Hinde, H. P. 1954. The vertical distribution of salt marsh phanerogams in relation to tide levels. Ecological Monographs 24: 209-225.

Mason, H. L. 1949. Evidence for the genetic submergence of *Pinus remorata. In* Genetics, paleontology, and evolution, G. L. Jepsen, editor. Princeton Univ. Press, Princeton.

———. 1955. Do we want sugar pine? Sierra Club Bulletin 40: 40-44.

———. 1957. A flora of the marshes of California. Univ. Calif. Press, Berkeley and Los Angeles.

McMillan, C. 1956. The edaphic restriction of *Cupressus* and *Pinus* in the coast ranges of central California. Ecological Monographs 26: 177-212.

Oosting, H. J., and W. D. Billings. 1943. The red fir forest of the Sierra Nevada. Ecological Monographs 13: 259-274.

Purer, E. A. 1942. Plant ecology of the coastal salt marshlands of San Diego, California. Ecological Monographs 12: 81-111.

St. Andre, G., H. A. Mooney, and R. O. Wright. 1965. The pinyon woodland zone in the White Mountains of California. American Midland Naturalist 73: 225-239.

Stevenson, R. E., and K. O. Emory. 1958. Marshlands at Newport Bay, California. Allan Hancock Foundation, Occasional Paper No. 20.

Storer, T. I., and R. L. Usinger. 1963. Sierra Nevada natural history. Univ. Calif. Press, Berkeley and Los Angeles.

Sweeney, J. R. 1956. Responses of vegetation to fire. Univ. Calif. Publications in Botany 28: 143-250.

Wright, R. D., and H. A. Mooney. 1965. Substrate-oriented distribution of bristlecone pine in the White Mountains of California. American Midland Naturalist 73: 257-284.

Chapter 6

Clements, F. E. 1936. The origins of the desert climax and climate. *In* Essays in geobotany in honor of W. A. Setchell, T. H. Goodspeed, editor. Univ. Calif. Press, Berkeley.

Jaeger, E. C., and A. C. Smith. 1966. Introduction to the natural history of southern California. Univ. Calif. Press, Berkeley and Los Angeles. Natural History Guide Series.

Munz, P. A. 1962. California desert wildflowers. Univ. Calif. Press, Berkeley.

Went, F. W. 1955. The ecology of desert plants. Scientific American, April.

Chapter 7

Axelrod, D. I. 1958. Evolution of the Madro-Tertiary geoflora. Botanical Review 24: 433-509.

Campbell, D. H., and I. L. Wiggins. 1947. Origins of the flora of California. Stanford Univ. Publications in Biological Science 10: 3-20.

Munz, P. A. 1959. A California flora. Univ. Calif. Press, Berkeley and Los Angeles. pp. 5-10.

Robbins, Wilfred W., Margaret K. Bellue, and Walter S. Ball, 1951. Weeds of California. State of California, Sacramento.

II. General references

The following list of publications and comments has been selected largely from Helen K. Sharsmith's 1967 version of "An annotated reference list of the native plants, weeds, and some of the ornamental plants of California" and an unpublished revision compiled by Alice Q. Howard. Ultimately, this should be available from the Agricultural Extension Service, University of California, Berkeley CA, 94720.

Abrams, LeRoy. 1923, 1944, 1951, 1960. Illustrated flora of the Pacific states. Stanford Univ. Press, Stanford. 4 volumes. A comprehensive technical flora of the native plants of the Pacific coastal states, every species illustrated.

Armstrong, Margaret. 1915. Field book of western wild flowers. G. P. Putnam's Sons, New York. Over many years this has been a very popular layman's field book to the more common wild flowers found west of the Rocky Mountains.

Arnberger, Leslie P., and Jeanne R. Janish. 1968. Flowers of the southwest mountains. Southwestern Monuments Assoc., Globe Ariz. This layman's handbook includes a considerable number of arid California species.

Baerg, Harry J. 1955. How to know the western trees. Wm. C. Brown Co., Dubuque, Iowa. A popular guide to the native and cultivated trees of the Rocky Mountains westward.

Baker, Richard St. B. 1960. The redwoods. Naturegraph Co., Healdsburg, California. Interesting stories about the Coast Redwood and the Sierra Big Tree.

Bakker, Elna S. 1971. An island called California. Univ. Calif. Press, Berkeley and Los Angeles. An introduction to natural plant and animal communities of the state.

Balls, Edward K. 1962. Early uses of California plants. Univ. of Calif. Press, Berkeley and Los Angeles. Natural History Guide Series. An account of the uses to which indigenous California plants have been put by Indians, Spaniards, pioneers, and present-day inhabitants, for food, fiber, medicine, etc.

Benson, Lyman. 1969. The native cacti of California. Stanford Univ. Press, Stanford. Designed for use by the botanist and layman alike. A detailed study of the botany and taxonomy of every species, variety, and major hybrid growing native or naturalized in California including keys, descriptions, ecology, and physiology.

Benson, Lyman, and Robert A. Darrow. 1954. The trees and shrubs of the southwestern deserts. Univ. New Mexico, Albuquerque, and Univ. Arizona, Tucson. Semipopular or semitechnical manual for identification of trees and shrubs of the deserts of southwestern United States.

Berry, James B. 1924. Western forest trees. Dover Publications. New York. A guide to the identification of trees and woods for students, teachers, farmers, and woodsmen, arranged according to leaf characteristics.

Boughey, A. S. 1968. A checklist of Orange County flowering plants. Museum of Systematic Biology, Univ. Calif., Irvine.

Bowerman, Mary L. 1944. The flowering plants and ferns of Mount Diablo. Gillick Press, Berkeley. A floristic and ecological study of the native plants of Mount Diablo, Contra Costa County, useful to both the scientist and the layman.

Bowers, Nathan A. 1965. Cone-bearing trees of the Pacific coast. Pacific Books, Palo Alto. A popular descriptive account of the native coniferous trees of the Pacific states, the species grouped according to altitude and geography.

Brockman, C. Frank. 1947. Broad-leaved trees of Yosemite National Park. Yosemite Natural History Assn., Yosemite Nat. Park. A pocket guide to the non-coniferous trees of Yosemite.

Clements, Edith S. 1927. Wild flowers of the west. National Geographic Magazine. Over 200 of the more conspicuous wildflowers of the Pacific Coast described and illustrated by color paintings by the author.

———. 1959. Flowers of coast and sierra. Hafner Co., New York. Popular account of the more conspicuous or well-known plants of the Pacific Coast from Washington to California. Identification mainly by colored figures, descriptions meager.

Cole, James E. 1939. Cone-bearing trees of Yosemite. Yosemite Natural History Assn., Yosemite Nat. Park. A pocket guide to the coniferous trees of Yosemite.

Collins, Barbara J. 1972. Key to coastal and chaparral flowering plants of Southern California. Calif. State Univ. Foundation, Northridge. Includes more than 475 plants, simplified key with illustrated glossary of terms.

Cook, Lawrence F. 1961. The giant sequoias of California. U.S. Govt. Printing Office, Washington. A short descriptive account of the Big Trees of the Sierra Nevada, their distribution, geological and natural history, conservation, etc.

Cooke, William B. 1940. Flora of Mount Shasta. American Midland Naturalist Volume 23. The Univ. Press, Notre Dame, Indiana. 1941. First supplement to the flora of Mount Shasta. Volume 26. 1949. Second supplement to the flora of Mount Shasta. Volume 41. The introduction contains information on history, physiography, geology, climate, streams, and life zones. All ferns, conifers, and flowering plants are keyed and briefly treated as to location.

Dawson, E. Yale. 1966. The cacti of California. Univ. Calif. Press, Berkeley and Los Angeles. Natural History Guide Series. A popular illustrated guide to the cacti of the state.

Dodge, Natt N., and Jeanne R. Janish. 1969. Flowers of the southwest deserts. Southwestern Monuments Assoc., Globe, Ariz. Handbook for the layman on the more common and conspicuous plants of the southwest deserts; includes a considerable number of California species.

Engbeck, Joseph H., Jr. 1973. The enduring giants. Univ. Calif. Extension, Calaveras Grove Assn., Save-the-Redwoods League, Calif. Division of Parks & Recreation. The natural history of Sierra Big Tree, including paleobotany.

Ferris, Roxana S. 1962. Death Valley wildflowers. Death Valley Natural History Assn., Death Valley. Popular handbook of the most frequently met plants of Death Valley National Monument, each species illustrated with clarity and charm. Identifications by means of illustrations, flower color, habit, and habitat, amplified by interesting descriptions.

———. 1968. Native shrubs of the San Francisco Bay region. Univ. Calif. Press, Berkeley and Los Angeles. Natural History Guide Series. A popular illustrated guide to shrubs around San Francisco Bay, but also useful in much of northern California.

———. 1970. Flowers of the Point Reyes National Seashore. Univ. Calif. Press, Berkeley and Los Angeles. Black & white drawings of all species. Arranged by family with a color index additionally.

Fultz, Francis M. 1923. The elfin forest. Times-Mirror Press, Los Angeles. Classic work on the chaparral.

Geary, Ida. 1972. The leaf book. Tamal Land Press, Fairfax. More than 350 scientifically accurate prints made from the plants themselves. Includes seaweeds; fungi, lichens, & mosses; ferns and fern allies; grasses, sedges, & rushes; flowers; shrubs; and trees.

Gillett, George W., John Thomas Howell, and Hans Leschke. 1961. A flora of Lassen Volcanic National Park, California. Wasmann Journal of Biology, Volume 19, University of San Francisco, San Francisco. A technically excellent treatment of this local flora, also of value to the layman. Species are keyed first to family, then to genus and species. In the annotated catalogue, localities and habitats are given for each species.

Grillos, S. J. 1966. Ferns and fern allies of California. Univ. Calif. Press, Berkeley and Los Angeles. Natural History Guide Series. An introduction to the ferns and their allies most commonly encountered in the field.

Hall, Harvey M., and Carlotta C. Hall. 1912. A Yosemite flora. Paul Elder & Co., San Francisco. An excellent book on the flora of Yosemite for the botanist and advanced amateur, still very useful despite its early date of publication.

Hood, Mary, and Bill Hood. 1969. Yosemite wildflowers and their stories. Flying Spur Press, Yosemite. Contents are selected from among those with showy flowers. Black and white photos, line drawings, key, and map. Historical and ecological notes included. Indices to common names, scientific names, and to persons and mythology.

Hoover, Robert F. 1970. The vascular plants of San Luis Obispo County, California. Univ. Calif. Press, Berkeley and Los Angeles. A flora approach with occurrence, keys to species, distributions, and descriptions included.

Howell, John T. 1970. Marin flora. Univ. Calif. Press, Berkeley and Los Angeles. Manual of the flowering plants and ferns of Marin County, California. A technical manual useful to both botanist and advanced amateur, which is also aesthetically very satisfying. Its beautifully written introductory chapters appeal to all Marin County enthusiasts.

Howell, John T., Peter H. Raven, and Peter Rubtzoff. 1958. A flora of San Francisco, California. Wasmann Journal of Biology, Volume 16, Univ. San Francisco, San Francisco. An account of one of our outstanding local floras which is rapidly being

being depleted by urbanization. Annotated catalogue of the species.

Howitt, Beatrice F., and John T. Howell. 1964. The vascular plants of Monterey County, California. Univ. San Francisco, San Francisco. An annotated catalogue of the flowering plants and ferns of Monterey County.

Jaeger, Edmund C. 1968. Desert wild flowers. Stanford Univ. Press, Stanford. Desert plants arranged by families, with habitat, range, flower color, etc., without keys and mostly without descriptions, but copiously illustrated. Includes both Mojave and Colorado deserts.

Jepson, Willis L. 1910. The silva of California. Univ. Calif. Press, Berkeley. A monumental work on the trees of California.

———. 1909-1922, 1936, 1939, 1943. A flora of California. Associated Students Store, Berkeley, and California School Book Depository, San Francisco. 3 volumes, unfinished. An amplification of the technical account of the seed-plants of California as presented in Jepson's "Manual of the flowering plants of California". Documented; with geographical, biological, and bibliographic notes.

———. 1923. The trees of California. Associated Students Store, Berkeley. Discussion of geographic ranges, outstanding characteristics, uses, and other interesting facts about California's native trees, together with an annotated list of the species.

———. 1923-25. A manual of the flowering plants of California. Univ. Calif. Press, Berkeley and Los Angeles. The first technical flora of the entire state, with keys and descriptions to all the native and introduced seed plants and ferns of California known in 1925. The classic account of the botany of California.

———. 1935. A high school flora for California. Associated Students Store, Berkeley. A reissue of the author's previous "Flora of the economic plants of California". A simplified technical treatment of the more common native and cultivated plants.

———. 1935. Trees, shrubs and flowers of the redwood region. Save-the-Redwoods League, San Francisco.

Kirk, D. 1970. Wild edible plants of the western United States. Naturegraph Publishers, Healdsburg, California. Describes or mentions about 2,000 species of higher plants, the bulk of the edible plants of the region, and includes methods of preparation. Poisonous plants are also described.

Legg, K. 1970. Lake Tahoe wildflowers, and of the central Sierras. Naturegraph Publishers, Healdsburg, California. Describes the commoner and more beautiful wildflowers of this area.

Lenz, Lee W. 1956. Native plants for California gardens. Rancho Santa Ana Botanic Garden, Claremont, California. Gardeners wishing to grow California natives will find here a wealth of information on the characteristics, propagation, and horticultural uses of native species recommended on the basis of the long experience in growing them at the Rancho Santa Ana Botanic Garden.

Lindsay, George. 1963. Cacti of San Diego County. Society of Natural History, San Diego. A beautifully written popular booklet which will delight those interested in cacti.

Lloyd, Robert, and Richard S. Mitchell. 1973. Plants of the White Mountains, California and Nevada. Univ. Calif. Press, Berkeley and Los Angeles. Botanical history, discussion of plant communities and vegetation, phytogeography and comparative floristics, geology. Listing of the vascular plants with keys to their identification.

McClintock, Elizabeth, and Walter Knight, with Neil Fahy. 1968. A flora of the San Bruno Mountains, San Mateo County, California. Calif. Academy Sciences, San Francisco. Annotated plant list with discussion of history, topography, geology, climate and vegetation.

McMinn, Howard E. 1939. An illustrated manual of California shrubs. Univ. Calif. Press, Berkeley. Technical treatment of about 800 species of native shrubs with keys and semi-popular descriptions, well-illustrated.

McMinn. Howard E., and Evelyn Maino. 1937. An illustrated manual of Pacific Coast trees. Univ. Calif. Press, Berkeley. A manual of all Pacific Coast trees, both native and horticulturally grown, of use to both the botanist and the interested amateur. Simple keys, and well-drawn botanical descriptions for each species.

Metcalf, Woodbridge. 1959. Native trees of the San Francisco
    Bay region. Univ. Calif. Press, Berkeley and Los Angeles.
    Natural History Guide Series. Characteristics, ranges, habitats,
    and relation to man of all native trees of the bay region.
Muller, Katherine K. 1958. Wildflowers of the Santa Barbara
    region. Santa Barbara Botanic Garden, Inc., Santa Barbara. A
    beautiful little booklet describing and picturing 58 wildflowers
    of the Santa Barbara area.
Munz, Philip A. 1961. California spring wildflowers. Univ. Calif.
    Press, Berkeley. Popular account of the more frequent spring
    wildflowers, arranged by flower color. Every species is illus-
    trated by line drawings or kodachromes.
———. 1963. California mountain wildflowers. Univ.
    Calif. Press, Berkeley. Mostly summer wildflowers of the
    mountainous areas, with a few spring and autumn species.
———. 1964. Shore wildflowers of California, Oregon, and Wash-
    ington. Univ. Calif. Press, Berkeley. The fourth book in Munz'
    popular series, this contains plants of the Pacific shoreline,
    "that portion of the coast which is influenced by salt spray".
———. 1973. A California flora and supplement. (In collaboration
    with David D. Keck). Univ. Calif. Press, Berkeley and Los
    Angeles. Comprehensive technical treatment of all known spe-
    cies of ferns and seed plants of California. Introduction con-
    tains a section on geological history by D. Axelrod, and a dis-
    cussion by Munz and Keck of the 29 plant communities they
    recognize for the state.
———. 1974. A flora of Southern California. Univ. Calif. Press,
    Berkeley and Los Angeles. The standard text and reference
    book on the flowering plants of southern California.
Niehaus, Theodore. 1974. Sierra wildflowers: Mount Lassen to
    Kern County. Univ. Calif. Press, Berkeley. Natural History
    Guide Series. The nearly 550 wildflowers described in this
    guide represent mostly the common ones found in the several
    plant communities of the Sierra Nevada.
Parsons, Mary E. 1966. The wildflowers of California. Dover
    Publications Inc., New York. Short popular descriptions of
    the common or conspicuous California flowering plants, group-
    ed according to flower color, with key to genera. This book,

first issued in 1897, has long been a favorite.

Peattie, Donald C. 1953. A natural history of western trees. Houghton Mifflin Co., Boston. All trees native to western North America are included in this magnificent book. Each is described historically, aesthetically, botanically, and geographically, and each is illustrated with a woodcut. A key to all species and a glossary is given at the end of the book.

Peñalosa, Javier. 1963. A flora of the Tiburon peninsula, Marin County, California. Wasmann Journal of Biology, Volume 21, Univ. San Francisco, San Francisco. A well-written account and annotated checklist of the small but very interesting and diversified flora of the Tiburon peninsula.

Peterson, P. Victor. 1966. Native trees of southern California. Univ. Calif. Press, Berkeley and Los Angeles. Natural History Guide Series. A guide to aid in the recognition of the native trees only, with descriptions of the leaves, fruit, and flowers where applicable. Includes distribution, interesting characteristics, and a short general description.

Peterson, P. Victor, and P. Victor Peterson, Jr. 1974. Native trees of the Sierra Nevada. Univ. Calif. Press. Berkeley and Los Angeles. Natural History Guide Series. A popular illustrated guide to trees in the Sierra Nevada.

Philbrick, R. N. 1972. Plants of Santa Barbara Island. Madroño, Volume 21, California Botanical Society Inc., Berkeley. An annotated list of the vascular plant flora with a discussion.

Raven, Peter H. 1963. A flora of San Clemente Island, California. Aliso, Volume 5, Rancho Santa Ana Botanic Garden, Claremont.

———. 1966. Native shrubs of southern California. Univ. Calif. Press, Berkeley and Los Angeles. Natural History Guide Series. A popular illustrated guide to native shrubs of southern California.

Rickett, H. W. 1970. Wild flowers of the United States. Volume 4: The southwestern states. McGraw-Hill, New York. Covers about 3,000 species in New Mexico, Arizona, Nevada, and southern California as far north as the Transverse Ranges and the deserts to approximately the latitude of Mount Whitney. Limited to herbaceous wildflowers, omitting those with inconspicuous flowers (and therefore no trees, shrubs, grasses, sedges, rushes), but including some cacti despite their woody nature. Color photographs for most taxa included.

———. 1971. Wild flowers of the United States. Volume 5: The northwestern states. McGraw-Hill, New York. Covers, within the same limitations as the previous listing, about 3,000 species in Washington and Oregon west of the Cascade Range, and California south to the deserts. Again, color photographs for most taxa included.

Rodin, Robert J. 1960. Ferns of the Sierra. Yosemite Natural History Assn., Yosemite. A detailed semitechnical presentation of all ferns of the Sierra, with keys and photographs, together with a discussion on ferns and fern allies, their story and life history.

Rubtzoff, Peter. 1953. A phytogeographical analysis of the Pitkin Marsh (Sonoma County). Wasmann Journal of Biology, Volume 11, Univ. San Francisco, San Francisco. A geographically oriented study of the vegetation and flora of the Pitkin Marsh "floral island" in Sonoma County, with an annotated list of the species occurring there.

Sharsmith, Helen K. 1945. Flora of the Mount Hamilton Range of California. American Midland Naturalist, Volume 34, The Univ. Press, Notre Dame, Indiana. A taxonomic study and floristic analysis of the vascular plants of the area. Physiography, geology, climate, soils, plant communities, etc., followed by an annotated list of species.

———. 1965. Spring wildflowers of the San Francisco Bay region. Univ. Calif. Press, Berkeley and Los Angeles. Natural History Guide Series. Over 300 species of wildflowers found in the nine counties touching San Francisco Bay, of which about 215 species are keyed, briefly described, and illustrated. Introductory chapters acquaint the amateur with the hows, whys, and wheres of wildflower study.

Smith, A. C. 1959. Introduction to the natural history of the San Francisco Bay region. Univ. Calif. Press, Berkeley and Los Angeles. Natural History Guide Series. A general introduction to the natural history of the San Francisco area covering physical features, climate, seasons, and informative sections on the plants and animals.

Smith, Gladys L. 1962. Flowers of Lassen. Loomis Museum Assn., Mineral, Calif. A booklet for the visitor to Lassen Volcanic National Park, the more conspicuous flowering plants of the park illustrated and described.

———. 1963. Flowers and ferns of Muir Woods. Muir Woods Natural History Assn., Mill Valley, Calif. This is a booklet for the visitor to carry with him on the trails of Muir Woods.

Sudworth, George B. 1908. Forest trees of the Pacific slope. Dover Publications Inc., New York. Descriptive account of all trees of the Pacific coast, each species illustrated and with a full statement of its range. No keys. Nomenclature modernized by E. S. Harrar.

Sweet, Muriel. 1962. Common edible and useful plants of the west. Naturegraph Co., Healdsburg, Calif. Ferns, water plants, vines, trees, shrubs, and herbs which have been or are useful are illustrated and described briefly. Poisonous plants are carefully designated. An interestingly written account for the layman.

Thomas, John H. 1961. Flora of the Santa Cruz Mountains of California. Stanford Univ. Press, Stanford. A technically detailed and authoritative treatment of the plants of the Santa Cruz Mountains, of value to the botanist and advanced amateur. Individual species are carefully keyed, and habitats and localities are given.

Thomas, John H., and Dennis R. Parnell. 1974. Native shrubs of the Sierra Nevada. Univ. Calif. Press, Berkeley and Los Angeles. Natural History Guide Series. An illustrated guide to the native shrubs of the Sierra Nevada.

Thorne, Robert F. 1967. A flora of Santa Catalina Island, California. Aliso, Volume 6, Rancho Santa Ana Botanic Garden, Claremont, Calif. This work contains an annotated checklist of the plant species and a short study of the ecological characteristics of the area.

Thurston, Carl. 1936. Wildflowers of southern California. Esto Publishing Co., Pasadena. Popular treatment of almost all the seed plants and ferns of southern California from Santa Barbara county southward. Well illustrated with 547 photographs.

True, Gordon H. 1973. The ferns and seed plants of Nevada County, California. California Academy of Sciences, San Francisco. A listing with brief introductory essay, index to families and to localities.

Twisselmann, Ernest C. 1956. A flora of the Temblor Range. Wasmann Journal of Biology, Volume 14, Univ. San Francisco, San Francisco. A well-written and fascinating account of the flora of this unit of the Diablo Range, its ecology, history, etc.

———. 1967. A flora of Kern County, California. Wasmann Journal of Biology, Volume 25, Univ. San Francisco, San Francisco. An annotated check list of the plants preceded by a lengthy and excellent discussion of natural history, geography, geology and soils, climate and weather, floristics, plant associations, numerical analysis of the flora, centers of distribution and origin of the flora, range limits in Kern county, and botanical exploration in the area.

Watts, Tom. 1963. California tree finder. Nature Study Guild, Berkeley. A handy little pocket manual for identification by the amateur of the more common native trees of California.

Williams, J. C., and H. C. Monroe. 1969. The natural history of the San Francisco Bay area. McCutchan Pub. Corp., Berkeley. Designed to foster an understanding of the natural history of the area. Contains numerous illustrations and photographs of the biota.

Witham, Helen V. 1972. Ferns of San Diego County. Natural History Museum, San Diego. Traces the history of ferns, their ancient origin, uses and beliefs in the Middle Ages, then lists the 26 ferns of San Diego county and helps the reader to identify them by drawings, photographs, and descriptions.

INDEX TO COMMON NAMES

143

144

145

INDEX TO SCIENTIFIC NAMES

149

lobata, 65, 97
turbinella, 67, 106
vacciniifolia, 66
wislizenii, 64, 65, 97

Rafinesquea, 7
Ranunculus californicus, 77
Rhamnus californica, 63
Rhododendron, 7
  macrophyllum, 63, 86, 88
Rhus diversiloba, 63, 64, 68, 93,
  109
  integrifolia, 58, 68, 109
  laurina, 58, 64, 93
  ovata, 64, 93
Ribes, 66, 67, 102, 104
Romneya coulteri, 58
Rosa, 9
Rubus, 8, 9, 102
  parviflorus, 8, 58, 64, 86
  vitifolius, 62
Rumex crispus, 124

Salazaria mexicana, 56, 69, 111
Salicornia, 62, 68, 79, 110,
  Plate 5D
Salix, 65, 68, 79, 100, Plate 15D
Salvia, 65, 67, 100, 104, 112
  leucophylla, 68, 108
  mellifera, 68, 108
Sanicula arctopoides, 62, 78
Sarcobatus vermiculatus, 29, 45,
  68, 110
Saxifraga mertensiana, 59
Schmaltzia, 7
Scirpus, 65, 100
  pumilus, 25
Scoliopus bigelovii, 59
Scrophularia, 7
Senecio clevelandii, 17
Sequoia, 4, 87, 115
Seqoiadendron giganteum, 19, 58,
  66, 101, 116, Plate 11D
Sequoia langsdorfii, 89, 90
  sempervirens, 4, 6, 18, 28, 63,

69, 85, 87-90, Plate 7B
Sidalcea hickmanii, 121
Simmondsia chinensis, 8
Sisyrinchium, 11
  bellum, 77
Sonchus asper, 124
Sorghum halepense, 123
Spartina foliosa, 79
Stipa, 65, 98
Streptanthus, 27
  campestris, 58
Suaeda, 56, 68, 79, 110
  californica, 62

Tamarix, 123
Taxodium, 6, 88
  sempervirens, 6
Tellima, 8
Tetradymia, 68, 107, 111
  axillaris, 69
Thamnosa montana, 56
Thuja, 115
Thuja plicata, 63, 85
Torreya, 7
Triglochin, 79
Trillium, 11
  ovatum, 59
Tsuga, 4, 9, 115
  heterophylla, 63, 85
  mertensiana, 67, 104, Plate
  12C
Typha, 28, 65, 100

Ulex europaeus, 125
Ulmus, 115
Umbellularia, 4, 89
  californica, 3, 4, 64, 65, 86,
  88, 89, 121, Plate 8A
Urospermum picroides, 124

Vaccinium, 67, 104
  ovatum, 59, 63, 64, 70, 86,
  88
Vancouveria parviflora, 59, 70,
  88

151